Remarkable Engineers

Engineering transformed the world completely between the seventeenth and twenty-first centuries. *Remarkable Engineers* tells the stories of 51 of the key pioneers in this transformation, from the designers and builders of the world's railways, bridges and aeroplanes, to the founders of the modern electronics and communications revolutions. The focus throughout is on their varied life stories, and engineering and scientific detail is kept to a minimum. Engineer profiles are organized chronologically, inviting readers with an interest in engineering to follow the path by which these remarkable engineers utterly changed our lives.

IOAN JAMES is Emeritus Professor of Mathematics at the University of Oxford and has had a distinguished career as a research mathematician. In recent years he has become interested in the history and development of scientific disciplines and the scientists involved. He was elected Fellow of the Royal Society in 1968. He is also the author of *Remarkable Mathematicians, Remarkable Physicists* and *Remarkable Biologists*, all published by Cambridge University Press.

Remarkable Engineers

From Riquet to Shannon

Ioan James

University of Oxford

Shaftesbury Road, Cambridge CB2 8EA, United Kingdom

One Liberty Plaza, 20th Floor, New York, NY 10006, USA

477 Williamstown Road, Port Melbourne, VIC 3207, Australia

314–321, 3rd Floor, Plot 3, Splendor Forum, Jasola District Centre, New Delhi – 110025, India

103 Penang Road, #05–06/07, Visioncrest Commercial, Singapore 238467

Cambridge University Press is part of Cambridge University Press & Assessment, a department of the University of Cambridge.

We share the University's mission to contribute to society through the pursuit of education, learning and research at the highest international levels of excellence.

www.cambridge.org
Information on this title: www.cambridge.org/9780521516211

First published 2010
Reprinted 2013

A catalogue record for this publication is available from the British Library

ISBN 978-0-521-51621-1 Hardback
ISBN 978-0-521-73165-2 Paperback

Contents

Preface

This work is intended for those who would like to read something, but not too much, about the life stories of some of the most remarkable engineers born since the Renaissance. There are five or six profiles in each of nine chapters, making 51 engineers altogether. The emphasis is mainly on their varied life stories, not so much on the details of their achievements. Although I knew none of them personally – most of them died long before I was born – I know something of their works. In France I have sailed along Riquet's Grand Canal de Languedoc, been impressed by the fortifications of Vauban and ascended the Eiffel tower. In England, I have seen mighty beam engines at work, and in museums. I have ridden on the footplates of steam engines, and I have frequently used Brunel's Great Western Railway. In the United States, I have walked across Roebling's Brooklyn Bridge and have inspected the Wrights' biplane in the Smithsonian National Air and Space Museum. In the Second World War, I had first-hand experience of the V-1 flying bomb and the V-2 ballistic missile. In Russia, I have flown in one of Tupolev's aircraft. My house is full of electrical appliances, as is the car I drive. I write this book on a laptop computer, the descendant of Babbage's analytical engine, which was to be powered by steam. Nowadays we live in a world dependent on the work of generations of engineers.

Prologue

Most people know what an engineer is without being able to produce a definition. We say that someone has engineered the solution to a problem, and the dictionary allows this by defining an engineer as someone who contrives, designs or invents, with the same root as genius, a word whose meaning has varied much over the years. This covers not only traditional types of engineering, building bridges or railways, for example, or cars or aeroplanes, but also modern types, such as software engineering. Engineering overlaps with science, on the one hand, and with technology, on the other. There are many specialities: civil (as opposed to military) engineering, mechanical, electrical, medical, sanitary, computer, etc., are in common use. Feibleman (1961) has attempted to distinguish between these, but the distinctions matter little for my purposes. I give several examples of people who might be classified as applied physicists, others who might be regarded as electrical technologists, but they are still engineers. Although I have written about some of these engineers before (James, 2004; 2009a; b), the profiles here are not the same.

The profiles that follow are arranged chronologically by date of birth, so that when read in sequence they convey in human terms something of the way in which engineering developed. In writing this book, I had in mind the reader who, like myself, is interested in engineering but is not necessarily familiar with the history of the subject. To avoid being too discursive, I have focused in this book on certain themes. At first the emphasis is on civil engineering, the building of fortifications, canals, bridges, tunnels and so on. Then comes the development of the steam engine and the railway age. Electrical engineering became increasingly important, leading to the development of radio and television. The automobile dates from the early twentieth century, as does aviation. Finally, there is information technology and space research. Even so, certain forms of engineering had to be left out if this book was to be of a reasonable length. Some profiles are much shorter than others, mainly due to the lack of suitable biographical material.

Although I begin with two examples from the seventeenth century, most of my remarkable engineers date from the eighteenth, nineteenth and twentieth. It is hard to identify engineers, however remarkable, before the time of the Renaissance. We know the names of the architects of some of the outstanding buildings of the Middle Ages, but not much else. Bridges, fortifications, docks and canals were built, major drainage schemes were successfully completed, but we know little about the people who were responsible. Polymaths like Leonardo da Vinci and Michelangelo Buonarroti attracted the attention of biographers but these were exceptional. From the time of the Renaissance, more information is available, particularly for architects. Architecture used to be regarded as part of engineering, or vice versa, but engineering as a profession is a later development. In France, for example, the two professions of architect and engineer, which had previously been quite close, began to separate in the eighteenth century.

In pre-Revolutionary France, the centralized state controlled entry into the profession. The Corps Royal du Génie in 1817 was responsible for fortifications while the similar Corps des Ponts et Chaussées, which was responsible for the maintenance of highways, included several architects among its members. To ensure a good supply of civil engineers, the Ecole des Ponts et Chaussées was founded in 1747, while the Ecole de l'Artillerie et du Génie in Mézières, founded the following year, trained military engineers. These rival corps were often in competition with each other in the eighteenth century, especially over the desire of engineers belonging to the Ponts et Chaussées to work on canals and on the planning of docks, which traditionally had been the prerogative of those in the Génie. Entry to these military organizations was more or less restricted to members of the nobility. After the revolution, the Ecole Polytechnique was founded, to which entrance was not so restricted. Other countries followed suit, for example, the German Technische Hochschulen and the British polytechnics were modelled on the Ecole Polytechnique.

Until the eighteenth century, Britain was a backward country in engineering, compared with, say, France or the Netherlands. The Industrial Revolution reversed the situation. Britain led the way with the invention of the steam engine and the construction of railways. Having been trained on the job, in a business context, the British engineers were more empirical and more alert to questions of profit than were the French, while the French engineers attained a deeper mastery than their British counterparts of questions having to do with the rational exposition of technical knowledge. Telford and Rennie learned French in order to read treatises that had not been translated into English. France exported expert engineers like Marc Brunel and Charles Vignoles, some of whom settled in Britain. Over a dozen British engineers are profiled in this book but it would be easy to find as many more with equal claim.

In the early days of the Republic, the United States was primarily an agricultural country, but after the Civil War it began to surpass Britain in certain industries, and enjoyed a golden age of invention. In the 40 years or so before 1914, the advance in technology slowed down in Britain while it speeded up in America and in Germany. There was a technological lag in both the old industries and in the new but other factors were also important (see Habakkuk (1962), for example).

In Britain, and to some extent in the United States, it was left to the budding engineer to find his own way of entering the profession, usually by serving some kind of apprenticeship. This survival of the medieval guild system was the well-recognized means for the training of mechanical engineers, especially. A young man, very rarely a woman, would seek a master, who would probably test him in some way. He would sign articles, a set of rules, and for a period of years would learn a trade, being paid little or nothing. Usually, a good master might have several apprentices at a time. Traditionally, when an apprentice had served his time he stood a treat for the others. He might then travel around, picking up work wherever he could. Eventually, he would settle down and practise his trade in one place. Those who could afford it might pay a premium for their training, usually in a leading firm. However, engineers were also trained in military schools: in Britain there was the Woolwich Academy and in the United States, West Point, but this was less important than in France.

I have chosen subjects from a wide range of countries, namely Croatia, France, Germany, Hungary, Ireland, Italy, Russia, Scotland, Sweden, the Ukraine and the United States. Many of them migrated from the country of their birth to another country where they spent the major part of their career. Thus, Watt, Rennie and Baird moved to England from Scotland, the senior Brunel from France, Marconi from Italy, Gabor from Hungary; Roebling moved to the United States from Germany, Ericsson from Sweden, Bell from Scotland, Tesla from Croatia, Zworykin from Russia and von Braun from Germany. Oriental countries are not represented, mainly due to the lack of suitable biographical material. I have no doubt that there were remarkable Chinese, Indian and Japanese engineers, for example. In the case of China, we are fortunate to be able to consult the multi-volume *History of Science and Civilisation in China*, written by Joseph Needham (1954–2004) and his associates, but not much is on record about the individuals who were responsible for the advances in technology, which in many cases later spread to the West.

Women were excluded from the professions until relatively recently: engineering was no exception, rather an extreme case. As a result, it was difficult to find suitable subjects to profile. The problem is both the shortage of women engineers and the lack of sufficient biographical information about those we know of. In the end, I chose the British electrical engineer Hertha Ayrton (1854–1923) and the American electrical engineer Edith Clarke (1883–1958), but there were certainly others. Mechanical engineering does not seem an obvious vocation for a woman, and yet Victoria Drummond (1897–1980) made a successful career as a marine engineer, vividly described in her biography (Drummond, 1994). In nineteenth-century America, we find the sanitary engineer Ellen Swallow Richards (1842–1911) and the industrial production engineer Lillian Gilbreth (1878–1972). There were many cases where the wife contributed to the husband's work. Even where she worked independently, as did the wives of William Ayrton and Lee de Forrest, there was a tendency to attribute what she achieved to her husband. There is not the same shortage of women inventors – windscreen wipers, laser printers and bullet-proof vests were all invented by women – but few of them would be described as engineers.

A special issue of the tenth volume of the journal *History and Technology* describes the situation as it was in a number of different countries. Sally Hacker (1990), in *Doing it the Hard Way*, is mainly concerned with the situation in America at the end of the twentieth century. She maintains that the need for mathematical knowledge is an important factor, but this seems doubtful. Some knowledge has always been required but until fairly recently this could be learnt at school. Today, engineering mathematics is a highly technical subject but specialists are available for consultation. Moreover, many studies have shown that, broadly speaking, women are just as good at mathematics as men. However, Hacker is surely right when she says that engineering, like other professions, has a strong male tradition, which is not easily overcome. She quotes research that shows that in present-day America engineering contains the smallest proportion of females of all the major professions, and projects a heavily masculine image, which is hostile to women. Women are not easily accepted as colleagues by men in the workforce.

The position of black people in the United States was anomalous, to put it mildly. By the 1790s, there were nearly 700,000 Africans in the country. Some had been born free, others had earned their freedom fighting in the revolutionary war, and still others had bought freedom for themselves and their families. But 90 per cent were still held captive. In his youth, Thomas Jefferson stated, '[blacks] are much inferior, as I think one could scarcely be found capable of tracing and comprehending the investigations of Euclid,' although he changed his opinion later. In New Orleans, freemen were not permitted to send their children to the public schools and had to get permission to walk in the streets of the city. The State of Louisiana required both free and enslaved African-Americans to carry passes; the United States Supreme Court decided that blacks had no civic rights. Before the Civil War, a slave was not a person, just personal property of his or her master, and could not receive a patent for any discovery or invention. Nevertheless, slaves invented devices to make their work more efficient: for example, Eli Whitney's famous cotton gin was a development of such devices made by slaves for cleaning cotton.

One of the problems in the selection of subjects is that engineers seldom work alone. The Wright brothers seem inseparable, and so they share a profile. Several engineers, for example Rennie, Stephenson, Brunel and Sperry, had sons who carried on their work. There was usually a team of assistants, some of whom performed important tasks. Nowadays, there is almost always a firm to provide essential services. Another problem is that the person who is usually given the credit is not an engineer. The Suez Canal, for example, is surely one of the great engineering feats of the nineteenth or any century. Ferdinand de Lesseps is given the credit for it but he was a professional diplomat and businessman, not at all an engineer, as his subsequent failure in the case of the Panama Canal makes all too clear. In my first profile, which follows, the situation is not so clear-cut. Riquet was responsible for the Grand Canal de Languedoc, financed it and planned the route, but it was the Italian-trained engineer who constructed it, and others were involved as well.

1 From Riquet to Watt

Pierre-Paul Riquet (1604–1680)

The idea of a canal between the Atlantic and the Mediterranean, cutting out the long and dangerous haul around the coasts of the Iberian Peninsula, had been conceived by Leonardo da Vinci. Although much discussed, it remained no more than an idea until the middle of the seventeenth century, when the Grand Canal de Languedoc came into being. Voltaire, writing of the building achievements of the reign of Louis XIV, described it as 'le monument le plus glorieux' and Skempton describes it in the *History of Technology* (Singer *et al.*, 1954–84) as, 'the greatest feat of civil engineering in Europe between Roman times and the nineteenth century.' Nowadays the canal, known as the Canal du Midi, is mainly used by pleasure boats but there is still some commercial traffic.

The man who was responsible for the construction of the canal was born at the town of Béziers, not far from Montpellier, on 29 June 1604 to Guillaume Riquet, a wealthy lawyer, and his wife. The Riquet family are said to be of Italian origin, but centuries earlier they had settled in the Languedoc region of France. Riquet was educated at the Jesuit college in Béziers, where he excelled in science and mathematics, but he received no formal training in engineering. At 19, he married Catherine de Milhau, the daughter of a wealthy bourgeois family of Béziers, whose dowry was such that he was able to purchase the old chateau and estate of Bonrepos, near the little village of Verfeil 12 miles to the east of Toulouse, on the slopes of the valley of the river Girou. He thereby became Baron de Bonrepos, but I will continue to call him Riquet.

Seven years later, in 1630, Riquet was appointed as a collector of the salt tax in Languedoc, and it was not long before he became farmer general of the tax for the entire province. This tax was first introduced in France in 1206 and had become one of the chief sources of state revenue. The franchise was a lucrative but onerous one, which entailed a great deal of travelling throughout Languedoc and in this way Riquet gained an intimate knowledge of the country through which the canal would pass. In 1632, he appointed as his deputy an ex-school friend of his, named Paul Mas, a doctor of law who

was the cleverest lawyer in Béziers, and later became his brother-in-law. In the same year, Riquet's father died and his son inherited the principal part of his estate. He made the family house in the Place St Félix in Béziers the centre for his business in south-eastern Languedoc. This was extremely profitable and, assisted by the efficient Paul Mas, he made so much money that he was able to undertake the even more lucrative business of a military contractor, supplying the King's armies in Cerdagne and Rousillon, with the result that by the time he was 50 Riquet had amassed a fortune of several million livres. He acquired a town house in the Place St Pantaléon in Toulouse; and as the years went by he spent an increasing amount of time there.

The King, Louis XIV, and his great finance minister Jean-Baptiste Colbert (1619–83), had decided that the time had come to build the Grand Canal de Languedoc. They placed Riquet in charge of the project, ably assisted by a young engineer named Francois Andreossy (1633–88) and an official named Pierre Campas, well known to Riquet. Andreossy was a citizen of Narbonne but was born and educated in Paris, where he studied science and engineering. Having completed his education in the capital he returned to Narbonne, but in 1660 he went to Italy, where he had family connections, so that he could study the canals of Lombardy. Andreossy became Riquet's right-hand man throughout the period of the construction of the canal, supplying the professional expertise that Riquet lacked. It is debatable whether he or Riquet should be given the major share of the credit for the success of the great enterprise. However, it was Riquet, already 60 by the time construction began, whose drive and enthusiasm were essential. Riquet enlisted the support of the powerful Archbishop of Toulouse and together they travelled to Paris in 1662 to see Colbert. They explained to him the route of the canal and how they proposed to overcome the difficulties it presented. To canalize the natural channel of a river by means of locks and cuts was one thing; to construct a purely artificial waterway through high and difficult country and to ensure that it had an adequate water supply was quite another. Colbert was and remained enthusiastic and the king, on his advice, approved it, and appointed a royal commission to supervise the undertaking.

There were many unexpected difficulties, and at one stage the commissioners deemed it impossible to bore a tunnel through a sandy hill. By using a picked labour corps, Riquet managed to drill the tunnel before the commissioners heard about it, and when it was completed he lit the interior with triumphal torches and invited his critics to join him in a walk through it. After that he generally had his own way but the canal was not opened until 1681, after some 15 years of concentrated labour. More than 8,000 men were employed on the project. It is carried for considerable distances on built-up embankments, and 100 locks raise it 620 feet above sea level. The broad deep waterway is borne over rivers and through hills by many aqueducts, cuttings and tunnels. Although the Languedoc waterway followed the bed of the river Garonne for some of its length, on the canal section, between Agde and Toulouse, it was completely independent of the river, being fed by water brought from distant streams.

Riquet staked his private fortune on the vast undertaking and accumulated a huge debt. He never lived to see the canal complete and open to traffic. He was growing old and suffered from bouts of ill health as a result of endless work and worry. On 1 October 1680 he died at the age of 76. The canal was officially opened on 15 May the next year. Riquet died so encumbered by debts amounting to more than two million livres that his heirs were obliged to sell most of his shares in the concern. The canal did not begin to pay its way until 1724, some 43 years after it had been built.

An edict had been issued whereby the builders of the canal and their successors were to be granted the ownership of the canal in perpetuity with exemption from taxes on the property and various other valuable rights. This legal document was the subject of much argument when the king died in 1715. It was argued that huge sums of money had been expended on its construction by the king and the provincial parliament. In 1768, by which time the canal had become reasonably profitable, the States of Languedoc offered to purchase the canal for 8,500,900 livres but, although the purchase was approved by the king, the provincial government changed its mind and the canal remained in the hands of the shareholders until the outbreak of the French revolution. After the restoration, ownership was restored to the Riquet family until 1897, when it was taken over by the state, which abolished the tolls.

Sébastien le Prestre de Vauban (1633–1707)

The future exponent of siege warfare was born in the village of Saint Léger-de-Foucherest, south-east of Avallon in the Nivernais. The precise date of his birth is uncertain but it is known that he was baptised on 15 May 1633. His father, an impecunious country gentleman, sent him to the Carmelite college of Semur-en-Auxois to be educated. After spending seven years there he joined the rebellion of the Fronde as a cadet under Condé in 1651, but within two years had been taken prisoner. He then transferred his service to the King's side: he rapidly rose to be ingénieur ordinaire du roi, in which capacity he distinguished himself in the seven years of war against Spain that ended in 1659. The next year he married Jeanne d'Osnay, who came of a similar family to his own. He spent the next seven years partly in his birthplace of Saint Léger-de-Foucherest, but mainly travelling France in discharge of his military duties. He received the support and friendship of the Marquis de Louvois, Louis XIV's powerful minister of war, for whom he drew up in 1669 his *Memoir pour servir a l'instruction dans la conduite des sièges* (Vauban, 1740). In 1672, the king declared war on Holland, and, two years later, on Spain again. When his army was about to launch an attack, Louis XIV used to bring his court to watch the spectacle while enjoying a picnic.

In 1678, Vauban was appointed Commissary General of Fortifications, in which capacity he ringed France with fortresses, some of which are still intact while others can still be traced clearly from the air. Excellent photographs of the surviving fortifications will be found in Bornecque (1984). For defence, Vauban designed vast polygonal works of stone and earth, the

walls being defensible by flanking fire from musketry as well as artillery on the ramparts; further protection was provided by carefully sited two-storeyed gun chambers, whose thick walls and roof prevented the cannon from being put out of action. Vauban conducted a whole series of successful siege operations against fortresses guarding the Flemish frontier, culminating with the capture of the stronghold of Old Breisach in just a fortnight in 1703, the year in which he was created a Marshal of France. For attacking fortresses he elaborated the use of trenches dug as parallels to bring the besieging forces within striking distance, first at Maastricht in 1673, and invented the ricochet battery, which was first employed at Philipsburg in 1688. In the seventeenth century, war consisted above all in a series of sieges. It was said, not without truth, that once a town was besieged by Vauban, the town was taken. It was also said that a town fortified by Vauban was a town impenetrable, but his contribution to defensive tactics was less remarkable. The king created him Marquis d'Humières for his services.

Vauban was a military engineer, not an architect; the massive simplicity of his mighty walls was magnificent but the elaborate entrance gateways he built were amateurish and heavy-handed. Although fortifications were his speciality, Vauban was called upon to design the three major aqueducts required for Riquet's Languedoc canal and for the misconceived Maintenon canal, which was intended to convey the waters of the Eure to the chateau of Versailles, although this was never completed. Peace was declared in 1684, but four years later war broke out again and this time the French army suffered a series of major defeats. Although the country was virtually bankrupt, the king was squandering huge sums on Versailles and other projects. Vauban's attempts to influence the French government were unsuccessful. The autocratic and selfish king had surrounded himself with self-serving courtiers who would not listen to Vauban. For example, he made a splendid plea for the recall of the protestant Huguenots, who left France after the revocation of the Edict of Nantes in 1685. Most notable, however, were the statistical studies, which grew out of the surveys he made of so many regions for military purposes, posthumously published under the ironical title *Les Oisivetés* (leisure affairs). These provided the background to his famous protest against the system of privilege that was ruining France, *Projet d'une dixme royale* – a plea for a single progressive tax to be imposed on all classes. However, too many people benefited from the existing system of taxation, which made the rich richer and the poor poorer, and Louis XIV gave orders for the book to be suppressed. After the great marshal died of pneumonia in Paris on 30 March 1707, his funeral was attended by just a few friends and relatives. The ungrateful monarch he served so devotedly for over 50 years was not represented.

In 1699, Vauban had been elected an honorary member of the Académie Royale des Sciences and in 1705 was made a Chevalier de l'Ordre de Saint Esprit. His wife died the same year; she had seen little of her hard-working husband for most of her married life, and it seems likely that he was unfaithful to her. She had two daughters by him and one son who died in infancy.

James Brindley (1716–1772)

In the eighteenth century, England had a fine network of roads but most of them were in an appalling state. Transport by sea or river was used wherever possible. On the continent, canals had been constructed in various locations, and the technology was well developed. Transport by canal barge was slow but sure. When sails could not be used barges were hauled by men

or by horses. Horses could haul a much heavier load on a waterway than they could on even a well-maintained road. England was slow to realize the advantages of canals. Such canals as existed followed the course of rivers and other natural waterways. The Duke of Bridgewater showed that they could be constructed almost anywhere. Like Riquet, he became heavily indebted as a result.

Francis Egerton, the third Duke of Bridgewater, inherited property in various parts of the country. The most important of his estates was Ashridge, west of London, but in the north there was Worsley, near Manchester. A consumptive weakling in his youth, he was not expected to achieve much. His guardians packed him off on the customary extravaganza known as the grand tour of European countries, especially France and Italy. He was fortunate in his 'bearleader', who moderated his behaviour and made sure that, as well as having a good time, he gained an understanding of the arts and the industries of the places they visited. On his tour he saw the Grand Canal of Languedoc, and other feats of engineering. In Rome, he started to build up a collection of paintings that was later to become of the finest in the country, when he took advantage of the sale of the collections of the French aristocracy after the Revolution. On his return to England, he dissipated much of his fortune at the gaming tables of London, but when he tired of that he retreated to Worsley, where his estate included some coalmines. These produced no more than a small income, and he was determined to increase it.

The duke's agent, John Gilbert, devised a plan for a canal on which the coal could be transported from the duke's mines to Manchester, where it would be sold. This would be a great improvement over transport by packhorse. The duke was enthusiastic about the project, and set about raising the necessary funds. However, an engineer was needed to design and build the canal, and this is where the subject of this profile comes in.

James Brindley was born in a remote hamlet near Buxton, Derby, the son of a once-prosperous yeoman, who had lost most of his money through gambling. At the age of 17, after some years working as a ploughboy, he was apprenticed to a wheelwright and millwright near Macclesfield. He was, to a large extent, self-taught, since the master and journeymen of the shop were an incompetent and drunken lot. In 1742, he was able to start up on his own in Leek, repairing and setting up machinery of many kinds, including the flint mills required by Josiah Wedgwood for his pottery. In 1758, he patented an unsuccessful attempt to improve the inefficient Newcomen steam engine, used to pump water out of coalmines, and was employed in a survey for an abortive project for a Mersey–Trent canal route.

Brindley's famous association with the Duke of Bridgewater began in the following year, when the duke was reconsidering his Worsley canal scheme with a view to avoiding a costly descent by locks down to the river Irwell and then up again the other side. Brindley proposed an aqueduct to carry the canal over the river. Such aqueducts were a novelty in Britain although not on the continent, as the duke would know. The scheme was completed in 1761; it was a great success and the owners of the land through which it passed, who had tried very hard to stop its construction, were dumbfounded.

Brindley was subsequently employed to make the extension of the Worsley canal to Runcorn in 1762–7. The duke's triumph initiated what became canal mania. The members of the Bridgewater–Gilbert–Brindley triumvirate were involved in many projects, both large and small. The duke was one of the main promoters, with Wedgwood, of Brindley's biggest achievement, the Grand Trunk Canal linking the Mersey and the Trent across Cheshire and Staffordshire. Although this canal was not completed until six years after Brindley's death, he found time to lay out six others, making a total of 365 miles of canal.

Experience taught Brindley many things, but the ability of the semiliterate workman to solve the problems of such major works as the Burton aqueduct and the Hardcastle tunnel was surely the result of exceptional intelligence. Though he held strong opinions and prejudices, he was full of original ideas. He spoke with a strong Derbyshire accent, impressing the parliamentary committee considering the plan for the duke's canal. He was already 50 when he married a 19-year-old girl named Anne Henshall, daughter of his clerk of works on the Grand Trunk Canal. The marriage took place in 1765; they lived in a pleasant old-fashioned manor house at Turnhurst, Staffordshire. He died there on 27 September 1772. He left two daughters; his widow, still young, married again.

John Smeaton (1724–1792)

The engineer and instrument-maker John Smeaton was born at Austhorpe Lodge near Leeds on 8 June 1724. The eldest son of the attorney William Smeaton by his wife Mary, he had a younger brother, who died in childhood, and a sister, Hannah, eight years his junior. John entered his father's office to learn law after leaving Leeds Grammar School at the age of 16. At home, he spent his leisure time in a well-equipped workshop, wrapped up in mechanics. He was encouraged by the famous instrument-maker Henry Hindley of York to make some tools and clocks. Perhaps in the hope of reviving his son's interests in the law, Smeaton's father sent him to continue his legal studies in London in 1742, but to no avail, for with his father's consent and the encouragement of scientists in London he decided to become an instrument-maker like Hindley.

In 1748, Smeaton opened a shop of his own in London, selling Hindley's work as well as his own. Contact with instrument-making in the north, where the emphasis was on precision, led Smeaton to attach importance to accurate measurement throughout his life. The vacuum pump he produced in 1749 was capable of achieving rarefactions several times better

than any other model. His previous time in London had already brought him into contact with members of the Royal Society, who recognized him as a young man of exceptional intelligence, and, on the strength of papers read before them in the period 1750–4, he was elected to the fellowship in 1753. Six years later he gave a series of lectures to the Royal Society on his research into the natural powers of wind and water, for which he received the prestigious Copley medal. There was uncertainty as to whether overshot or undershot water mills were more efficient: he settled it decisively in favour of the former. For windmills and sailing ships, he found that the Dutch had already found the most efficient designs.

Smeaton had travelled through the Low Countries in 1755 to view harbour and drainage works in what is now Belgium and Holland; his detailed account of what he saw was published by the Newcomen society in 1938. When he returned, he successfully undertook several minor engineering projects and prepared plans for a projected bridge over the Thames at Blackfriars, which was not built. Smeaton's first important engineering commission came in the same year when, through the Earl of Macclesfield, then president of the Royal Society, he was engaged to rebuild the Eddystone lighthouse near Plymouth, which had been destroyed by fire. This challenging task he completed in a masterly and innovative manner, creating a prototype for all subsequent lighthouses built in the open sea. The story of the erection of the lighthouse has often been told, most ably by Smeaton himself in his classic *Narrative of the Building, and Description of the Construction of the Edystone Lighthouse*, first published in London in 1791 (Smeaton, 1791). When the lighthouse had to be replaced by a new one after 120 years, the one he constructed was re-erected on Plymouth Hoe in its creator's memory.

In 1747, Smeaton had proposed marriage to a young lady from Leeds but due to parental opposition this did not come off. Nine years later, both his parents having died by this time, he married Ann Jenkinson of York. Soon afterwards he left London and returned to Austhorpe Lodge with his wife and their first surviving child, a baby daughter, Ann. Two other daughters followed: Mary in 1761 and Hannah in 1765, who died young.

In 1757, he was asked to prepare a scheme for making navigable the upper Calder in Yorkshire. He carried out a survey and, based at Austhorpe Lodge, acted as superintendent engineer during construction of the works. The successful completion of the Eddystone lighthouse in 1758 brought Smeaton widespread recognition. In the following decade, his practice as an engineer grew rapidly, and he emerged as the first fully professional engineer in Britain. His reputation rested on his ability to use judgement founded on experience to provide the best answer to a particular problem at the lowest price compatible with sound construction. In this period, he was engineer for the Forth and Clyde Canal, begun in 1768, and repeatedly consulted by the proprietors of the Carron ironworks. Other successful projects in Scotland culminated in the creation of Aberdeen harbour, completed in 1780.

Smeaton had a long-standing interest in steam engines. From 1768 to 1772, Smeaton spent much time at Austhorpe experimenting with a model engine in order to determine the causes of the poor performance of an engine made for the New River Company in 1767. His experimental method, in which he varied one component at a time, holding all the others constant, led to improvements in engine efficiency. One of his colliery engines was 25 per cent more efficient and more powerful than any in existence. Later his improvements were overshadowed by those of James Watt, whose profile follows. In 1776, he contributed another paper to the Royal Society, in which he displayed a deep understanding of the concepts of work and power. The next years found Smeaton in continual demand as a consulting engineer in work of all kinds, usually with great success. His one failure was a bridge at Hexham, destroyed by an exceptional flood two years after it was built.

With the emergence in Britain of a growing number of first-class civil engineers, Smeaton organized a dining club in London, where practitioners from all sides of the civil engineering profession could meet one another in a friendly atmosphere to discuss their work and all manner of topics. It was later called the Society of Civil Engineers but its members were generally known as Smeatonians. Smeaton suffered a period of ill health in 1783. After his wife died the following year, his eldest daughter Ann took charge of the household at Austhorpe. He remained very active until 1791, when he retired, planning to spend the last years of his life in writing his memoirs, but in September 1792 he suffered a stroke while walking in his garden and died six weeks later on 28 October, in his 69th year. He was buried in Whitkirk parish church where his daughters erected a tablet to him and to their mother. Two hundred years later, a memorial stone was placed in the floor of Westminster Abbey, the national pantheon, inscribed 'John Smeaton, Civil Engineer, 1724–1792.'

James Watt (1736–1819)

Some of the greatest British engineers came from Scotland, most notably James Watt, who was born at the port of Greenock, not far from Glasgow, on 19 January 1736. His father, of the same name, was a carpenter and shipwright, and also a general merchant and part-owner of several small ships. He was a man of substance, respected in the local community. His wife, Agnes Muirhead, came of a Glasgow family; one of her kinsmen occupied the chair of humanity at the ancient university. Their son, one of several children, most of whom died in infancy, was physically frail. Worried that he too might not survive, his mother looked after the delicate boy devotedly.

As a boy, James Watt junior spent a good deal of time in his father's workshop, developing the manual skills that were to be of lasting value to him. At the local grammar school he excelled in mathematics. After his mother died in 1753, he went to Glasgow and then to London in 1755. To learn the craft of mathematical instrument-making he spent a gruelling year of work in a shop in Cornhill. On his return to Glasgow, he was appointed mathematical instrument-maker to the university. In 1759 he went into

partnership with an architect named John Craig; their business of making mathematical and musical instruments thrived until Craig's death after five years left his partner in debt.

The following year, he married his cousin Margaret Miller, by whom he had four children; two daughters and a son survived infancy. The same year he was asked to repair a model of a Newcomen steam engine. He pondered over the deficiencies of this machine, and in May 1765 arrived at his famous idea of keeping the cylinder hot and leading the spent steam to a separate condenser. The development of this idea was taken up by the wealthy industrialist John Roebuck, who owned an ironworks at Kinneil, near Bo'ness on the Firth of Forth, but poor craftsmanship, particularly trouble with the piston packing, made progress painfully slow, leaving Watt very dissatisfied. Roebuck and Watt were granted a patent in 1769, the terms of which were so very broadly drawn that almost any invention to improve the performance of steam engines would infringe it.

With a wife and young children to support, Watt needed to earn a living. He sold the instrument-making business in Glasgow and for the next seven years his main occupation was canal surveying in Scotland, including surveying the route for the Caledonian Canal from Inverness to Fort William. Watt was also sent to London to give evidence to the parliamentary committee considering the proposal for a Forth and Clyde canal, which experience did not impress him.

On the way back, Watt spent some time in Birmingham where he met Erasmus Darwin and William Small, the moving spirit of the famous Lunar Society. When he confided the secret of his invention to them they were enthusiastic for its development and communicated this enthusiasm to their friend, the great entrepreneur Matthew Boulton. After meeting Watt, Boulton tried to negotiate with Roebuck but without success until 1773, when Roebuck became bankrupt and Boulton acquired the patent of Watt's engine the following year.

After Watt's first wife died in 1773, he moved to Birmingham, with his children Margaret and James. This was a turning point in his fortunes. In 1775, Boulton succeeded, against strong opposition, in obtaining an extension of Watt's patent until the end of the century, and to Scotland as well as England. The famous partnership of Boulton and Watt then began. It took another five years of development to produce reliable and economic engines, and the firm did not show a profit until the 1780s. From 1775 until the 1790s, Watt worked hard on the development of his engine and made several important improvements; the double-acting engine, the rotative engine with sun and planet mechanism, parallel motion, the governor, and the idea of measuring output in horsepower. He and Boulton made many journeys to Cornwall and other places where their engines were being installed. They trained up William Murdock to deal with the Cornish mineworkers on their behalf; he proved to be a tower of strength.

In 1776, Watt married again, to Ann McGregor of Glasgow, and brought her to a house he had bought called Regent's Place, Harper's Hill, just a short walk from Boulton's Soho works in Handsworth in Birmingham. She bore him another son, named James like his father. Watt took a prominent part in the scientific activities of the Lunar Society as a result of which he took more interest in chemistry and in 1783 he suggested in a paper submitted to the Royal Society that water was not a simple substance but a compound of hydrogen and oxygen. This led to the complicated water controversy. Watt was elected a fellow of the Royal Society of London in 1785, having been elected to that of Edinburgh the previous year. Glasgow University awarded him the honorary degree of Doctor of Laws. The Academie des Sciences elected him one of their eight foreign associates. He declined the offer of a baronetcy.

In 1790, Watt left Regent's Place, where he had spent the most active and successful years of his life. He moved to a larger house, Heathfield, designed by Samuel Wyatt (brother of the more famous James) and built on an estate of 40 acres, on Handsworth Heath. His second wife was devoted to him but she ruled the household with a rod of iron. Unhappily both children of the first marriage died of consumption: the daughter Janet at the age of 15, the son Gregory, who seemed to share his father's exceptional ability, at the age of 27.

As the eighteenth century drew to a close, Boulton and Watt ceased to defend their master patent and inventors such as Trevithick felt free to develop their ideas. In the 1790s, Boulton and Watt began to hand over their business to the next generation and a new firm, Boulton and Watt and Sons, was formed. By 1800, Watt had virtually retired, but he was a workaholic. He could have lived a life of leisure but he continued inventing to the end of his life. Earlier, Watt had devised a chemical method for copying documents, which was used in offices for many years; this business was managed by his eldest son. Up in the attic of Heathfield, he spent his time developing sculpturing machines, which could manufacture a copy of a portrait bust. One of his first inventions had been a machine for drawing three-dimensional objects.

In temperament, Watt was inclined to be despondent and pessimistic. Although he lived to be 83, he was a lifelong valetudinarian and hypochondriac, subject to black moods of depression and self-depreciation. Watt's invention of the improved steam engine was epoch-making in two senses: first, it gave man better sources of power and is the very foundation-stone of the modern technological world; second, it led to the idea of control over nature, as opposed to waiting for the operation of natural forces.

Watt died at Heathfield on 25 August 1819, after a short illness. The poet William Wordsworth paid him this tribute:

> I look upon him considering both the magnitude and the universality of his genius, as perhaps the most extraordinary man this country ever produced; he never sought to display, but was content to work in that quietness and humility, both of spirit and outward circumstances in which alone all that is truly great and good was ever done.

To mark his heroic status, which grew in the years after his death, a colossal monument to his memory by the sculptor Chantrey was paid for by public subscription and placed in Westminster Abbey in 1834.

2 From Jessop to Marc Isambard Brunel

William Jessop (1745–1814)

Among Smeaton's apprentices was a young man who became first his assistant, then his partner. William Jessop has been unfairly neglected; because not very much is known about his life, apart from his work. His parents, Josias and Elizabeth Jessop, had three other children: two younger sons and one daughter. William, the future engineer, was born on 23 January 1745 at Devonport, where his father was employed. When Smeaton arrived in Plymouth in 1756 to build the new Eddystone lighthouse, he placed Josias in charge of the workyard and they worked together until it was finished three years later. It was hardly surprising when Josias' son William, who was keen to be trained in engineering, was accepted by Smeaton as an apprentice and thus William learned the basics of theoretical and practical engineering at Austhorpe Lodge.

At the age of 27, Jessop was beginning to act as Smeaton's junior partner. His first major work was in Ireland, where he extricated the government from difficulties over the construction of the Grand Canal that links the Liffey at Dublin with the Shannon near Banagher. Under his capable aegis, the line westwards was resurveyed, the fine Leinster aqueduct was built over the Liffey at Sallins, and the canal was driven successfully across the Bog of Allen. In 1773, still under 30, he was elected a member of the Smeatonian society. Four years later, he married Sarah Sawyer of Haddlesey House at Birkin church; after a short time in Pontefract they made their home at Fairburn on the north side of the Aire between Ferrybridge and Castleford; their first child, John, was born in 1779; their second, Josias, two years later.

Jessop became involved with a firm that was to become the biggest and best-known firm of early canal contractors, but he decided not to purchase a partnership although he retained an interest in the firm for the rest of his life. Instead he became consultant for some major projects in the Midlands and south of England. The great outburst of canal investment and speculation that we call the canal mania began in 1789 and ended in

1796. Jessop, recognized as the premier engineer, was involved in various capacities.

Over a period of 20 years, Jessop engineered a navigation system over a large area of the north-east midlands of England. The trunk of this system was the river Trent. He also designed and supervised the building of three major canals. Jessop worked with Telford from 1793 until 1812. During this period, he was appointed general agent, engineer and architect to the Ellesmere Canal Company, which had been formed to connect the river Mersey with the river Dee at Chester and then with the river Severn at Shrewsbury. The route chosen involved carrying the canal across the deep valley at a great height by means of an aqueduct. When the Pontcysyllte Aqueduct was opened after ten years in 1805, it carried the canal in a cast-iron trough upon stone piers over a length of 1,000 feet and at a height of 127 feet above the river. This is considered to be one of the great engineering achievements of all time. When the railways began to be built Jessop was heavily involved, and Jessop was also responsible for major works in the London docks and at Bristol. He was also involved in drainage work in the Fens and in building bridges.

In 1783, the Jessop family moved from Fairburn to Newark, convenient for the work he was undertaking. In 1786, he was elected alderman and held the office of mayor of Newark in 1790–1 and 1803–4. He was not primarily an architect, nor a great bridge-builder, but a man whose special skill lay in handling water and earth. According to his contemporaries, his manners were simple; when disengaged from business, and in the company of intimate friends, he not infrequently displayed a playfulness of disposition, and a fund of entertaining anecdotes. Totally free of all envy and jealousy of professional rivalry, his proceedings were free of all pomp and mysticism, and persons of merit never failed in obtaining his friendship and encouragement.

In his last years, it is said, he suffered much from a form of paralysis. He died on 18 November 1814, two months before his 70th birthday, and was buried in Pentrich churchyard. As Hadfield and Skempton (1979) conclude: 'Too much evidence of Jessop's work seems to have disappeared in too many places for it to be possible to give a full account of certain episodes in his career; in particular the whole of his personal papers have vanished.'

Lazare Carnot (1753–1823)

We now return to the France of Louis XIV. Lazare Nicholas Marguérite Carnot was born on 13 May 1753 into a Burgundian family that occupied a leading position in its locality. His father was a lawyer and a notary in the town of Nolay, on the Côte d'Or not far from the city of Beaune. He

had served the powerful duc d'Aumont, who in return acted as a patron for his four sons, of which Lazare was the youngest. The boy's ambition was to enter the military engineering school at Mézières. In this, he was unsuccessful at his first attempt but after spending a period at one of the schools in Paris, which prepared candidates for entry, he passed the examination and in 1771 began the two-year training course for engineering officer-cadets. Among his teachers was the famous mathematician Gaspard Monge, but there is no indication that he recognized Lazare Carnot's exceptional ability.

After Carnot graduated at the beginning of 1773, he was commissioned as lieutenant and for ten years assigned to boring garrison duties in north-eastern France. These were interrupted by a three-year posting to Cherbourg, where he was involved in the construction of the great harbour. It was then that he began writing scientific papers, of which the most important was his *Essai sur les machines en général*. This was the first attempt to discuss, in a theoretical way, the optimum conditions for the operation of machines of every sort. Another paper, which won a prize from the Dijon Academy, was a eulogy of Vauban. Yet another was his *Reflexions sur la métaphysique du calcul infinitésimal*, entered for a competition set by the Berlin Academy; although this did not win a prize, it was published, translated into various languages, and editions were still appearing many years later. He also wrote a paper suggesting how to steer hot-air balloons, and thus make them more useful for military purposes. Finally, he married Marie Dupont, whose family came from Saint-Omer, near Calais.

When the French Revolution began, Lazare Carnot's brother Feulins had already entered politics. Lazare followed his example in 1791 and won a seat in the National Convention in which the future of France was debated. Before long, he was chosen as the chief commissioner sent to persuade the commanding officers of the Army of the Rhine to accept the decrees of the National Assembly, which was accomplished successfully. In the Assembly, Carnot voted for the execution of Louis XVI and the repression of royalists. The executive power of the state was then placed in the hands of the extremist Committee of Public Safety, which orchestrated the Jacobin Terror. Carnot was a leading member of the Committee, and of the Directorate that replaced it as the government of France in 1795. In his capacity as a Director, Carnot occupied rooms of the Petit Luxembourg, which became the home of his family.

The new Republic was threatened by rebellion, which he vigorously repressed. More difficult to deal with was the first attempt, by an external coalition, to restore the Bourbon monarchy. Carnot organized the victory over the allied armies, using his engineering skills. For example, balloons were used to observe the movements of the royalist forces, and a semaphore telegraphic system was installed to provide rapid communication between the front-line troops and the authorities in Paris. However, after Bonaparte became First Consul, Carnot's triumph was cut short by a coup d'état, which sent him into exile. He avoided the next two tempestuous years by living in Switzerland and Germany before returning to France as Minister of War under Bonaparte. Soon it became clear that the two men were quite incompatible and Carnot resigned his office.

For the next few years, Carnot resumed his scientific work. He had been elected to the Institut de France in 1796, expelled after the coup in 1797, and re-elected in 1800. He belonged to the first class, that of physical and mathematical sciences, which had replaced the old Académie Royale des Sciences. As well as attending lectures he worked with enthusiasm on committees concerned chiefly with the refereeing of papers but also assessing inventions, such as the American Robert Fulton's steam-powered paddleboat. Among his later publications, the most important is his *Principes fondamontaux de l'équilibre et du movement* and *Géométrie de position*, both of which appeared in 1803.

Although still living in Paris, he played no further part in public life until, in 1815, he rallied to the Emperor's cause during the Hundred Days, serving him briefly as Minister of the Interior. When Bonaparte was finally defeated and the restored Bourbon monarchy began reprisals against the leading republicans, he moved for safety to Magdeburg in Germany, where he died on 2 August 1823, at the age of 70. His scientific work was continued by his son, Sadi, whose profile follows later.

Thomas Telford (1757–1834)

Scotland produced not only great mechanical engineers, like James Watt, but also outstanding civil engineers, notably Thomas Telford. He was born in a cottage at Westerkirk in Dumfriesshire on 9 August 1757. His father, John Telford, a shepherd, had died six months previously, so that he was raised by his mother Janet, whose only child he was. As a child, he worked for local farmers, when not attending school, but when he was 13 he was apprenticed to a local stonemason. He also became interested in poetry. At the age of 24, he set off for London, having previously spent a year working in the New Town of Edinburgh.

At this stage Telford's ambition was to become an architect. He was armed with letters of introduction to his fellow countrymen, the architects

Robert Adam and Sir William Chambers; the latter gave him a job as a journeyman mason, working on the new Somerset House. He proved so capable that he was rapidly promoted to superintendent. When this work came to an end he was taken up by the wealthy and influential William Pulteney in 1783. Through Pulteney, Telford became superintendent of some important building work in Portsmouth dockyard. Next, he was commissioned to superintend the renovation of the derelict Shrewsbury castle, one of many properties owned by Pulteney.

At the beginning of 1787, Telford was appointed Surveyor of Public Works for Shropshire. The story is often told of how he was consulted about the leaking roof of the city church of St Chad. On inspecting the building he discovered large cracks and advised that it was liable to collapse at any moment. His advice was treated with ridicule but only the next morning the tower of the church fell down, demolishing most of the nave. After that, his reputation soared.

Telford's first opportunity to design a church himself was in Bridgnorth, where he produced a competent but hardly striking design for the new church of St Mary Magdalene. Another of his early designs was the octagonal church of St Michael at Madeley, Shropshire. However, ecclesiastical architecture was not his forte. At the beginning of 1788, he was appointed Surveyor of Bridges for Shropshire. He tells us in his autobiography that he was responsible for no less than 40 major road bridges between 1790 and 1796, some of which were crossings of the river Severn. They were not all built of stone. The first iron bridge in England had been erected in Coalbrookdale a few years previously and Telford soon proved himself a master of the structural use of the new material.

During this period, Telford was appointed engineer to the British Fishery Society, which was formed to develop the Highland fishing industry by constructing a series of small fishing ports. Since communication by road was virtually non-existent in the Highlands, a survey was necessary, which Telford undertook in 1799. He wrote two reports on what he found, the second of which made a deep impression on the Government. It was decided to start by constructing the Caledonian Canal, linking the east and west coasts of Scotland through the Great Glen. Telford was appointed chief engineer and, despite enormous difficulties, it was ready for use in 1822. It was never a commercial success. Projects of this magnitude involved employing a large team of skilled and unskilled labourers; Telford took great care for their safety.

Altogether, 920 miles of new roads were built in the Highlands under Telford's supervision, and 280 more were realigned and remade. Over a thousand new bridges were constructed, including crossings of the rivers Spey and Tay in Scotland, and the Beauly and Dee in England. In addition, there were various harbour works. In England, also, Telford was asked to report on the state of the roads, particularly those which led to the Irish ferries. Among his recommendations was the improvement of the Holyhead road, the main route to Ireland, by building a first-class highway and replacing the ferry across the Menai Straits with a bridge. There was also the need for another major bridge at Conway. In each case, Telford designed a suspension bridge. The Conway Bridge, the first of its kind in Britain, spanned 327 feet between the suspension towers; the Menai Bridge spanned 579 feet. Both are still functioning today. Not all the bridges that Telford designed were built, notably that for a replacement of old London Bridge by a single iron arch spanning 600 feet. By this time, railways were beginning to compete with canals. Telford was on the side of the canal companies, who were fighting a losing battle. Telford himself favoured horses rather than locomotives to haul trains on tramways, which fed the canal barges.

Telford travelled incessantly; when in London he used the Salopian Coffee House at Charing Cross as his headquarters. In 1821, he acquired a home of his own, at 24 Abingdon Street, opposite the Palace of Westminster, which soon became a centre for civil engineering. Telford had long been a fellow of the Royal Society of Edinburgh; now he was elected to the Royal Society of London. He received a Swedish order of knighthood for surveying the route of the Göta canal, linking the North Sea with the Baltic, and supervising its construction. This was completed in 1832, by which time Telford was 75. Although suffering from hearing loss, he regularly presided over the new Institution of Civil Engineers, successor of the earlier Society. This, and similar professional bodies, performed functions that elsewhere were controlled by the state. They also conferred qualifications and honoured exceptional achievement in the profession. When he died, two years later, on 2 September 1834, he was buried in Westminster Abbey. He never married.

The writer Robert Southey got to know Telford well when they toured the Scottish highlands together. He found Telford attractive personally and a good companion, saying:

> There is so much intelligence in his countenance, and so much frankness, kindness and hilarity about him, flowing from the never failing well-spring of a happy nature, that I was on cordial terms with him in five minutes.

However, there was another side to Telford, shown in his brief autobiography *Life of Thomas Telford, Civil Engineer, Written by Himself* (but heavily edited by his protégé J. Rickman). In this he seldom mentions his equals and superiors; notably he never mentions Jessop, who was the senior engineer on several major projects for which Telford claimed the credit, including the famous Pontcysyllte Aqueduct. The same thing happened in the case of the Caledonian Canal. In fact it appears that he went so far as to destroy evidence of Jessop's contribution.

John Rennie (1761–1821)

Thomas Telford's main rival, the civil engineer John Rennie, was born on 7 June 1761 at the farmstead of Phantassie, near the picturesque East Lothian

town of East Linton, midway between Haddington and Dunbar. He was the youngest of the nine children of the freehold farmer James Rennie and his wife Jean. His father died when he was five. His formal education began at Dunbar high school but from the age of 12 he was also trained by Andrew Meikle, the inventor of the threshing machine, known as 'the ingenious miller'. At 19, he enrolled for three years at Edinburgh University, paying his living and tuition expenses by taking on small mechanical jobs. Scottish universities at the time differed greatly from those of Oxford and Cambridge. They were much more universities of the people, where untutored boys were sent to train for the professions, including engineering. Thus, Rennie completed his education and started work as an engineer. Already the trustees of the county of Midlothian had commissioned him to build a bridge across the Water of Leith, about two miles west of Edinburgh on the Glasgow road.

A visit to Watt in Birmingham led to Rennie's employment for the installation of the machinery in the Albion flour-milling works in London, the first to have an all-iron plant, powered by steam. After this had been operating successfully for three years it was destroyed by fire; arson was suspected. Up to this time, Rennie had been basically a mechanical engineer, but now he concentrated his energies more on civil engineering, succeeding to the position held by Smeaton as a general canal-work consultant.

Rennie's first major bridge was over the river Tweed at Kelso. This, with its semi-elliptical arches separated by pairs of Doric columns, its boldly defined masonry, and its level roadway, anticipates the chief characteristics of his later Thames bridges. He was to build well over 60 major bridges in the course of his career. However, his greatest challenge was to design and build a replacement for the old lighthouse on the dreaded Inchcape Rock, near the mouth of the river Tay off the east coast of Scotland, where many ships had been wrecked. The able and engaging Robert Stevenson was engineer to the Commissioners for Northern Lights, but at this stage he was still young and inexperienced and so Rennie was called in as chief engineer to design the 200-foot lighthouse and made a number of personal visits to the site. Work was started in 1803 and completed in 1810; after a heroic struggle. Credit for the success of this should be shared between Rennie and Stevenson, who went on to build many more lighthouses in the course of his career.

Rennie was often consulted about harbour improvement schemes in the United Kingdom and elsewhere. In some of these, where he was directly involved, he made useful innovations. At Hull, for example, he introduced a steam-driven dredging machine to reduce the labour of clearing the dock site. When he constructed the London and East India docks, downstream of London Bridge, he provided cast-iron columns and roofs for warehouses and steam-driven cranes on the quays. In 1811 he began building at Plymouth the first large breakwater on the English side of the channel. The provision of a stable slope of large stone blocks resting on a gigantic mound of rubble was much interfered with by the south-westerly gales but by the time of its successful completion, in 1814, Rennie's judgement of its effect was proved correct.

At this time there was no bridge across the Thames in London apart from old London Bridge. This medieval construction, with its narrow roadway, decrepit arches and battered starlings, was due for replacement, but there was an obvious need for additional bridges. Smeaton's design for a bridge at Blackfriars was not used, as we know. Rennie prepared designs for three new Thames bridges, but only two were built. The first was originally called the Strand Bridge, but later Waterloo Bridge. In 1809, Rennie and William Jessop were asked to report on a proposal for this. They were not impressed by the design, and so Rennie was asked to produce a better one himself. The bridge he designed was spanned by unusually wide elliptical arches. Realizing that the replacement for London Bridge would create an increased scour upstream, he laid the foundations of his piers in cofferdams, thus providing it with an absolutely secure base. Granite was extensively used in the construction. Waterloo Bridge was generally considered the finest bridge in the world at that time and is still regarded as Rennie's masterpiece. It was opened in 1817 by the Prince Regent, who offered Rennie a knighthood, which he declined. Its demolition in 1935–6 to be replaced by a wider bridge was against strong opposition.

After this triumph Rennie designed another Thames bridge, at Vauxhall, which was never completed owing to lack of funds, and a third at Southwark. This was composed of three massive cast-iron arches, resting on exceptionally solid masonry piers, to resist the flow of the river at its deepest and narrowest. Constructed in 1814–19, his Southwark Bridge was also replaced by a wider bridge in the 1920s. Finally, there was Rennie's design for the replacement of London Bridge itself. He located the new bridge 239 yards upstream from the old one and returned to stone masonry in his design, providing a centre arch with a span of 150 feet and a 29 foot clearance at high tide. However Rennie's health had begun to decline, at least partly due to overwork. He died in London on 4 October 1821 and was buried in St Paul's Cathedral.

Rennie ranks highly as one of the great British bridge-builders of the nineteenth century, and one of the last to use stone masonry. He used stone where Telford might have used iron, and when Rennie used iron, as in Southwark Bridge, his designs are comparatively ponderous. Otherwise all his bridges are admirable architecturally. He collaborated with Telford occasionally, the last of the few works in which they were both associated being the drainage of part of the East Anglian fens. In the course of his work, Rennie made many useful inventions but never applied for patents. As a result, he never became wealthy, although he built up an extensive business as a manufacturer of engineering machinery. He had no intimate friends but used to socialize at the Society of Civil Engineers, which he dominated.

In 1789, Rennie had married Martha Ann Mackintosh, who bore him nine children of whom six survived to maturity. When she died in 1806, one of Rennie's unmarried sisters came from Scotland to keep house for the family but she found that the polluted London air was bad for her health and returned to Scotland. The two eldest sons George, born in 1791, and John, born in 1794, became engineers of distinction. The former was the more talented and became a successful mechanical engineer, although he was handicapped by being a cripple. It was the entrepreneurial John who took over their father's business, including the construction of New London Bridge to his father's design, for which he received a knighthood on its completion in 1831. When road widening threatened this, after the Second World War, the relevant material was sold to an American company, who created a replica in Arizona as the centrepiece of a theme park. John gave this description of his father:

> He was naturally of a quick irritable disposition, so that he felt it necessary to keep it under control and schooled himself accordingly. His personal appearance was very dignified and imposing. He was well over six feet tall and powerfully built; in his prime he could walk fifty miles in a day without fatigue and could easily lift three hundredweight with his little finger. His head was extremely fine and majestic with a broad oval open countenance, large expressive blue eyes, high developed forehead, prominent nose slightly curved with proportionate mouth and chin, and splendid luxurious auburn hair.

Sir Marc Isambard Brunel (1769–1849)

Several of the most remarkable British engineers were French-born or had French forebears. For example, Marc Isambard Brunel was born in the hamlet of Hacqueville in the rich Vexin plain of Normandy on 25 April 1769. He was the third child and second son of Jean Charles Brunel, a moderately prosperous farmer, and Maria Victoire Lefebre; who died when he was seven. Originally destined by his parents for the priesthood, the boy convinced his father that his talents were more practical rather than theological, and so he should become an engineer instead. He was educated at the College of Giscours, where he began training as a military officer, but then was sent to the seminary of Sainte Nicaise in Rouen. Next he served in the French navy, first as a cadet, then as an officer for six years from 1786, during which he visited many of France's possessions in the Caribbean. When he returned to Paris he found that the French revolution was in its third tumultuous year.

Since Brunel expressed his staunchly royalist sympathies rather openly, he was in danger of being arrested if he remained in France, and so he emigrated to the United States, where before long he found work as

a land surveyor. Soon he began to practise as a civil engineer and architect. Based on what he had learnt in France he entered a competition to design a new Capitol building for the United States Congress. Although his design was not accepted, a modified version was used for the Bowery Theatre in New York. After taking American citizenship he was appointed chief engineer of the city in 1796, where he was employed to erect an arsenal and cannon foundry. He also strengthened the defences of the channel between Staten Island and Long Island.

One evening, Brunel was dining with Alexander Hamilton when another émigré, recently arrived from England, told him of the supply difficulties that Pitt's government were experiencing in expanding the Royal Navy to a level appropriate for the struggle with the French for control of the seas. The Admiralty required about 100,000 pulley blocks for its ships every year, for example, and these were handmade by skilled workmen. Although mass production was by no means unknown at this time – for example, Joseph Bramah (1748–1814) was producing his patent locks by this method – Brunel saw that there was an opportunity for the mass production of pulley blocks, and came to England in 1799, with a view to exploiting his ideas.

Before leaving France, Brunel had fallen in love with his future wife, a young English woman named Sophia Kingdom, the daughter of a Plymouth naval contractor. They had kept in touch during the six years he was in America. For part of the time she was imprisoned in a convent by the Jacobins, but after her release she managed to return to England. Brunel proposed marriage, was accepted and the wedding took place in London, where their first child, named Sophia after her mother, was born.

Meanwhile the Inspector-General of Naval Works authorized Brunel to set up his machinery at Portsmouth dockyard. A total of 43 block-making machines were erected, several of which remained in service for a century and a half. Driven by a 30-horsepower steam engine, they converted elm logs into blocks ready for fitting and polishing, and reduced the labour required from 110 skilled to 10 unskilled workmen. The installation was in full operation by 1808; the Navy Board made difficulties about payment for Brunel's services but eventually they paid up. So that he could superintend the process the Brunel family moved to Portsea, near Portsmouth on the English Channel, where two other children, another daughter, Emma, and their only son, Isambard, were born.

Brunel had patented his design for block-making machinery; another invention he patented was the circular saw, although such saws were already in use in America. His sawmills at Battersea, Chatham and Woolwich were kept busy. Brunel also prepared another project of mass-production, making boots for the Army, worn at the battle of Waterloo, which were successfully marketed to civilians as well. He designed one of the early steamships and tried unsuccessfully to convince the conservative Navy Board that steam-powered tugs could tow men-of-war from harbour and provide them with an offing despite unfavourable winds and tides. He also constructed a tramway to facilitate the movement of timber around Chatham dockyard.

When the Brunels moved back to London, taking a house in Chelsea, they had the scientists Sir Humphry Davy and Michael Faraday as neighbours and through them Brunel became accepted as a member of the scientific fraternity of the metropolis. He was elected a fellow of the Royal Society in 1814, the first of many honours, and, 14 years later was to become the first foreign-born vice-president.

Among his many other interests, Brunel was one of those who were trying to improve the efficiency of steam engines. He designed a large engine for use in shipping, which was remarkably advanced for its period. At the same time, Davy and Faraday were studying the possibility of constructing a new type of engine, involving changing liquid carbonic dioxide into gas, which they thought might provide power much more cheaply than the steam engine. Brunel took up this idea, which he patented, and at first made good progress with the gas engine, as it was called, but eventually the high pressures involved led him to conclude that it was impracticable.

After the restoration of the monarchy in France, Brunel restored his connection with the land of his birth, tendering unsuccessfully for works to supply the city of Paris with pure water. Another abandoned project was to bridge the river Neva at St Petersburg by a span of 800 feet, while at Rouen his plans for a bridge to link the city with the island of Lacroix were rejected because Brunel was not a member of the government corps of engineers. In England, he foresaw the development of the decorative packaging industry, filing a patent application for a new form of tinfoil, which became a popular novelty, but his invention was promptly pirated.

Although a very industrious and innovative engineer, Marc Brunel was a simple and unworldly man, rather casual about the management of his various undertakings, so that rogues took advantage of his trusting nature. He and his wife were, for a time, imprisoned for debt, following the failure of the bank to which he had entrusted his capital. Fortunately, he had influential friends in high places and his creditors were paid by the government, for services rendered, so that the imprisonment did not last more than a few months. His designs for suspension bridges across rivers in the Ile de Bourbon (now called Réunion) were accepted by the French government. Brunel made a plan for a canal across the Isthmus of Panama, a suspension bridge across the Thames at Kingston, a subterranean aqueduct to supply Hampstead with Thames water from Hammersmith, and so on.

Upstream of London Bridge, Rennie was building new crossings of the Thames, as we have seen in the previous profile, but downstream, the width of the river made this impossible and a tunnel was perceived to be the solution. There had recently been an attempt to bore one between Rotherhithe on the south bank and Limehouse on the north, but this had only got as far as the exploratory stage when the workings were flooded and the project was abandoned. Now there was to be another attempt, this time from Rotherhithe to Wapping, closer to London Bridge than the previous one. Unfortunately, it was not realized that at this location gravel dredging had made deep holes in the bed of the river so that what seemed to be a straightforward project turned out to be extremely difficult. Brunel designed a three-tier cast-iron shield as a stand for the excavators, with screw jacks to press it against the working face and the completed masonry. Repeated inundations, as we shall read in the profile of Brunel's son, held up the work, which was not completed until 1843. Although the tunnel was wide enough for carriages, because of problems with the ramps leading down from street level, it was only used for pedestrian traffic.

Worn out by the work on the Thames tunnel, which absorbed much of the time and energy of his later years, Brunel handed over his engineering practice to his capable son Isambard Kingdom, whose profile occurs later. There was no interruption of the work, and no change of professional attitude. Problems were subject to the same scrupulous analysis; visions were imparted with the same bold conviction; details were designed with the same deft artistry; and works were carried forward with the same resolute devotion. In 1842, a stroke temporarily paralyzed Brunel's right side, and another stroke three years later left him permanently paralyzed on the right side. A friend gave us a glimpse of the old couple:

> I believe I was a good listener, and assured of my sympathy they poured out their reminiscences freely, or rather I should say Lady Brunel did so, for the old man was not voluble, though he often by a nod of the head or some short exclamation confirmed his wife's words. She was a little old lady, with her faculties bright and apparently unimpaired; he with a ponderous head surmounting what might be called a thickset figure. The old couple sat side by side and often the old man would take his wife's withered hand in his, sometimes raising it to his lips with the restrained fervour of a respectful lover.

He was a devoted husband and father, fond of society and a music lover. There existed between them a relationship of the most profound and mutually enriching kind. In his old age he wrote a touching tribute: 'To you, my dearest Sophia, I am indebted for all my successes.'

Although the son is more famous, the father is considered the more original engineer. Isambard supported his parents during their happy autumn years in a modest house overlooking St James Park. After his father died on 12 December 1849, at the age of 80, his mother moved to Duke

Street and spent her widowhood in a room with a balcony overlooking the park. After five lonely years she died on 5 January 1855, and was laid to rest at her husband's side. Rolt, in his famous biography of the son (2006), summarizes the father's character in the following words:

Like Brindley, Rennie, Telford and George Stephenson, Brunel was self-taught, a born craftsman with a flair for invention. He was an unself-conscious man, and had those qualities of simplicity, unworldliness and natural dignity which are the natural attributes of the craftsman in any age. From these qualities sprang an implicit trust in his fellow men which was so frequently misplaced. Hence his repeated disappointments; hence, too, the fact that he so often failed to secure an adequate return for his labours, although he received the richly deserved honour of a knighthood.

3 From Trevithick to Sadi Carnot

Richard Trevithick (1771–1833)

Throughout his life, the mechanical engineer Richard Trevithick was dogged by misfortune and never achieved worldly success, He was born at Pool in the parish of Illogan, Cornwall on 13 April 1771; he had four sisters, but no brothers. His father held the responsible position of manager at the important Dulcoath mine. While he was still a baby, the family moved to Penponds, just outside Camborne, where he went to school and learned the three R's. He grew up into an immensely strong young man, capable of lifting heavier weights than anyone else around. While still quite young he was appointed engineer, responsible for erecting and servicing the steam engines that pumped out water from the pits. In those days, the alternative to the Newcomen engine was the one supplied by the firm of Boulton and Watt.

In 1797, Trevithick's father died and his 26-year-old son became sufficiently well-off to marry. His bride was Jane Harvey, daughter of a local businessman. A woman of strong character who spared him domestic worries, she stood by him through thick and thin, despite the fact that with him work took precedence over family life. Trevithick was convinced of the need to increase the efficiency of steam engines to cope with the increasing depth of the pits but, like many others, was held back by Watt's master patent, covering every conceivable innovation, and Watt was opposed to the use of steam at much more than atmospheric pressure. When the patent finally expired in 1800, Trevithick built a double-acting high-pressure engine, with a system of diverting waste steam up the chimney to increase the draught, a simple but important innovation.

Trevithick installed a smaller version of his engine to power a steam carriage, which on Christmas Eve 1801 carried passengers over short distances over some extremely bad roads. Apart from engineering difficulties there was strong opposition to these early steam carriages because they frightened the horses, and so Trevithick turned his attention to railway engines.

Historians of technology have traced industrial railways back to Elizabethan times in England, and there were some in other countries. A number of lines were constructed in the first quarter of the nineteenth century, mostly quite short, but the longest was over 30 miles in length. Typically the loaded wagons were transported to the nearest waterway under gravity; on the return journey the empty wagons were drawn by horses. When steam locomotives became available, they were sometimes used instead of horses, although their weight caused damage to the track. Industrial railways like these carried passengers only to a very limited extent, if at all.

First Trevithick went to London armed with various letters of introduction. He was encouraged by his fellow Cornishman Humphry Davy and by Count Rumford to patent his versatile power unit. This was accomplished without too much difficulty. His aim, however, was to develop a locomotive that would run on rails, replacing the horses that were used to haul trains of wagons. His first attempt was at the Pen-y-Darren ironworks near Merthyr Tydfil, where he was employed as engineer. It could draw trains of wagons containing 10 tons of iron ore and 70 men. However, the 10-ton locomotive proved too much for the cast-iron rails of the tramway and the project was abandoned. He built another version, which ran for entertainment on a circular track in London, but although passengers were charged a shilling for a ride it did not pay. A high-pressure steam engine of the Trevithick type was built at Gateshead in 1805 but proved too heavy for the wood rails of the Wylam colliery. Two more were built, of which one made a brief appearance on Tyneside. Two Trevithick-type locomotives with pinion engaging with iron rack rails were successfully built in 1812 to carry coal from the Middleton collieries to Leeds. William Chapman of Durham also built one in 1813, which employed bogies to soften its impact on the rails, and in the same year William Hedley built another, but at Gateshead.

Although Trevithick was not involved in these later developments, nevertheless he had already established the major principles of locomotion. In particular, he refuted the current belief that friction between smooth iron wheels and rails would be insufficient to provide traction. Discouraged by repeated failures, Trevithick abandoned his efforts to establish steam traction and concentrated on finding other uses for his versatile power units as stationary engines. Nearly 50 engines had been built by 1804 and applied with varying degrees of success for pumping water, threshing and grinding corn, dredging, etc. Some were exported to the West Indies for use in sugar mills. Paddle steamers were already at an experimental stage; Trevithick designed one of his own, but although it operated successfully on a trial run, he was unable to interest anyone in developing it.

Trevithick could never resist a challenge. With his experience of tunnelling in the Cornish mines, he agreed to help construct the Rotherhithe–Limehouse tunnel under the Thames but, as we know, that was abandoned. Meanwhile, his financial situation was deteriorating. Instead of consolidating what he already had, he was pressing ahead with other new inventions, which he patented but for the development of which he was unable to secure finance. In 1811, after being forced to sell his shares in the patents for the high-pressure steam engine, he was adjudged bankrupt. After the bankruptcy, Trevithick had returned to Cornwall, and spent some time with his wife and four children, two daughters and two sons. In 1812, Jane gave birth to a third son, named Francis, who later wrote a biography of his father.

In Spanish America, this was a time of political upheaval, which culminated in the wars of independence. Opportunities opened up for British entrepreneurs to obtain mining concessions, and they consulted British engineers. Trevithick had a visitor from Peru, who told him of an exciting discovery of silver at Cerro de Pasco, high up in the Andes. This was crying out for mechanization and Trevithick agreed to supply the necessary machinery, broken down into parts that could be carried to the mine. Unfortunately, there was no-one capable of assembling the parts correctly and so, in 1816, Trevithick went to Peru himself. Before long, the mine was working well and he was fêted as a national hero. However, the miners left because they were afraid of being forced to serve in the army and when the rebel forces came to Cerro de Pasco they wrecked the machinery that Trevithick had installed. He was offered a contract to mechanize the Peruvian mint but nothing came of this.

A few years later Trevithick left Peru for more stable Costa Rica, where he invested in a gold-silver mine. The mine reached the stage when more finance was needed, and since this could not be obtained locally, he decided to return to England in the hope of finding investors there. He made his way down to the Atlantic coast, which was not far away, but there was not even a footpath and so he experienced a dangerous and difficult journey through the jungle, during which he lost most of his possessions. Arriving destitute at the Colombian port of Cartagena, he found Robert, the son of George Stephenson there. Robert Stephenson, who was also consulting on behalf of British investors, gave Trevithick £50 with which to buy his passage home. He reached Liverpool, by a roundabout route, in 1827. His family, who had not heard from him for many years, were naturally astonished when he turned up unannounced.

Further disappointments awaited him. Potential investors in Trevithick's Costa Rican mine wanted an independent assessment of its prospects before they would put up any money. Some of his patents had been earning during the years he was away but for one reason or another there was not much for him. He made some money by converting some of the Boulton and Watt steam engines to high pressure. Then he invented a new type of cannon for the Royal Navy; but the Admiralty did not take it up. A visit to Holland to inspect some pumping engines seemed about to lead to an attractive offer of employment but unfortunately he lost his temper at a critical point in the negotiations.

By this time Trevithick was in his 60s. In 1830, he went back to London, leaving his family in Cornwall. In declining health he spent the last years of his life working in Dartford, Kent. He proposed to build a 1,000-foot open-lattice cast-iron tower in London to commemorate the passage of the first Reform Act. The tower, resting on a 60-foot masonry plinth, was to be 100 feet in diameter at the base and 10 feet wide at the top, surmounted by a 49-foot wide platform carrying a gigantic 38-foot statue.

No practical progress had been made, either in finding the money or building the tower, when he died on 22 April 1833, not long after Eiffel was born. After his death, he was for many years a forgotten man but although unfairly overshadowed by George Stephenson, he now occupies an honoured place with other pioneers of the steam engine.

Sir George Cayley (1773–1857)

In 1909, Wilbur Wright, the American pioneer of aviation, wrote 'about a hundred years ago an Englishman, Sir George Cayley, carried the science of flying to a point it had never reached before and which it scarcely reached again during the last [nineteenth] century.' Cayley was born on 27 December 1773 at Scarborough on the Yorkshire coast. His mother, Isabella Seton, was the dominant parent; his father, the baronet Sir Thomas Cayley, suffered from ill health and spent much of his time abroad. Although George Cayley had four sisters, he was the only son. After a short time at school in York, he was sent to be educated by the brilliant non-conformist George Walker in

Nottingham. Walker, a fellow of the Royal Society, was not only well read in mathematics and navigation but was also a skilled mechanic. Cayley continued his education by studying under George Morgan, another able non-conformist scientist, under whose influence he became interested in the hot-air engine and in aerial navigation, two of the subjects to which he was to devote much of his time during the next 50 years of his life.

On the death of his father, at the age of 64, George Cayley succeeded to the baronetcy in 1792, and for the rest of his life occupied the old family mansion of Brompton Hall, between Pickering and Scarborough. Three years later he married Sarah Walker, daughter of his former tutor. Of highly uncertain temper, she had both beauty and brains; they had three sons and seven daughters, but three of the children died young. When he came of age, he also succeeded to the family's extensive estates in Yorkshire. While not neglecting the responsibilities of a Yorkshire squire, he devoted almost his whole life to study and experiment on the basic principles of flight. His first biographer called him the father of aerial navigation and further research has shown this claim is fully justified.

He first experimented with model gliders, using twisted rubber as a source of power. In 1849, Cayley constructed a small glider capable of carrying a small boy, and four years later a full-size one, which reputedly carried his coachman on a flight of some 500 yards. He understood the lift given by the cambered wing and distinguished this from drag. After some experiments with flapping-wing models (orthicopters) he recognized the advantage of fixed-wing design. He clearly formulated the concept of vertical tail surfaces, of the rudder for steering, of elevators and the airscrew. He was also interested in airships and was the first to suggest a semi-rigid vessel, with the gas enclosed in separate cells.

In his monograph of 1962, the aeronautical historian C. H. Gibbs-Smith has chronicled in detail the steps by which Cayley arrived at his conclusions.

> With imagination, perspicacity, determination and outstanding intelligence, he first of all clarified his ideas by formulating the problems involved in aviation; then he faced these problems from the strict standpoint of technology, and set about solving them. He quickly abandoned the pure orthicopter concept, and he divorced the system of lift from the system of thrust, thus introducing the concept of a flying machine with fixed main lifting surfaces, to be driven through the air by some auxiliary means of

> propulsion. Having arrived at the idea of a fixed wing to sustain the machine, he was next faced by the implicit problems of how to secure the stability and steerage of such an aircraft. Cayley therefore progressed in an orderly fashion through these problems, at first solving them theoretically (in essence); next incorporating his findings into models and finally applying them to full-size vehicles. He only stopped short of carrying out pilot-controlled gliding tests, which must have seemed unjustifiably hazardous in view of the construction difficulties he believed he faced at the time.

Practical powered flight was in any case impossible then, because of the lack of an engine with a high enough power-to-weight ratio. Steam engines were much too heavy but it seemed that an internal combustion engine, using heated air under pressure, would have the right characteristics. Cayley was one of a number of leading engineers who tried to construct such an engine but without much success. He invented the caterpillar tractor to facilitate transport over rough ground, and also designed artificial limbs for those who had suffered an amputation. He made several proposals for improving the safety of railway travel: early locomotives had inefficient brakes and there were frequent accidents.

In Yorkshire, Cayley was affectionately remembered as one of the most responsible resident gentry, known to every farmer and labourer from their childhoods. He initiated the system of agricultural allotments. He organized and presented a petition to parliament to allow a large-scale land drainage scheme in his part of Yorkshire. From his youth, Cayley was a keen proponent of parliamentary reform. He served as chairman of the powerful Whig club in York and, following the passing of the first Reform Act in 1832, was elected the Member of Parliament for Scarborough. With Wilberforce he played an active part in the movement to abolish slavery in the British Empire.

Cayley's social and professional circle included many of the great engineers and scientists of the late eighteenth and early nineteenth centuries, including Thomas Young, John Dalton, Sir Humphry Davy, Charles Babbage, George and Robert Stephenson, John Rennie, Sir Goldsworthy Gurney and the seventh Duke of Argyll. In 1831, he sponsored the first meeting, in York, of the British Association for the Advancement of Science, of which he became a life member. In 1836, he was elected a Fellow of the Royal Society. In education, Cayley is remembered as a founder of the Polytechnic Institution of Regent Street, London, where lectures and demonstrations were given to amuse and educate the public, and to provide evening classes in science, engineering, navigation, and even the principles of driving rail ay locomotives. He died at Brompton Hall on 15 December 1857, and was succeeded in the baronetcy by his only surviving son, Digby, born in 1807.

Cayley's posthumous reputation suffered almost complete neglect until the end of the century, and even today his full stature is not generally recognized. Gibbs-Smith (1962) has collected English, French and American tributes to his work including the following, written by the French aeronautical historian Charles Dolfuss in 1923 and translated by Gibbs-Smith.

> The aeroplane is a British invention: it was conceived in all essentials by George Cayley, the great British engineer who worked in the first half of the last century. The name of Cayley is little known, even in his own country, and there are very few who know the work of this admirable man, the greatest genius of aviation. A study of his publications fills one with absolute admiration both for his inventiveness, and for his logic and common sense. This great engineer, during the Second Empire, did in fact not only invent the aeroplane entire, as it now exists, but he realized that the problem of aviation had to be divided between theoretical research – Cayley made the first aerodynamic experiments for aeronautical purposes – and practical tests, equally in the case of the glider as of the powered aeroplane.

George Stephenson (1781–1848)

George Stephenson, the future railway pioneer was born at Wylam, near Newcastle upon Tyne, on 9 June 1781. The son of Robert Stephenson, a colliery fireman, he began in his father's trade. At 17, he was put in charge of a new pumping engine at Water Row, where his father worked. He moved to other pits at Walbottle before going to Black Callerton as brakeman in charge of the winding engine. In 1802, he moved to service a similar engine, installed to draw ballast wagons from Willington Quay to the top of a dump, and finally to Killingworth West Moor pit in 1804.

It was while he was working at Black Callerton that he married Frances Henderson on 28 September 1802. She bore him a son, Robert, who will be profiled later, in 1803, and a daughter in 1805, who did not survive. Frances herself died of consumption the following year, at the age

of 37. When his father was injured in a boiler house accident, which left him blind and unable to work, Stephenson seriously considered following the example of his recently married sister Anne and emigrating to America but could not afford the fare.

At Killingworth, through caring for the education of his son Robert, he acquired a new interest in life. He rose steadily in the service of the 'Grand Allies', who controlled the pits. He was put in charge of the machinery at all their collieries in 1812, and was allowed to work for other colliery owners as well, provided he continued to keep their own plant functioning properly. During this period he built some 39 stationary engines – one of 200 horse power – and was responsible for rationalizing coal handling above and below ground by replacing horse-drawn sleds with trains running on rails and drawn by stationary engines. Amongst other mining improvements, he devised a safety lamp in 1815, for which he was thanked three years later with a gift of £1,000 raised by public subscriptions. This brought him into competition with Sir Humphry Davy, who had devised a similar lamp about the same time.

The mounting cost of horse fodder prompted colliery owners to experiment with steam-driven locomotives running on cast iron rails rather than wood. Stephenson's employers commissioned him to build a locomotive for Killingworth at their West Moor workshops. The *Blücher* took to the rails in July 1814: it embodied the important innovation of running on flanged wheels; previously locomotives had smooth wheels designed for running on flanged rails or ordinary roads. Stephenson also introduced steam-cushioned suspension for smoother running. As a result of this, and other improvements, he was invited to join the Walker Ironworks in 1815 on a part-time basis to exploit his inventions, which included the steam blast. In 1820, Stephenson married again, to Elizabeth Hindmarsh, the daughter of the largest farmer in the parish of Black Callerton, whom he had known since childhood.

Cut off from easy access to the river Tees at Stockton, the Bishop Auckland coalfield was severely handicapped in comparison to that on the Tyne. Schemes for linking it by canal had been discussed since 1768 but nothing was done until 1821, when an Act of Parliament approved the laying of a railway between Stockton and Darlington. As well as building steam locomotives and other machinery, Stephenson had also gained some experience of laying out industrial railways, building an eight-mile one for the Thetson colliery. When he was appointed engineer for the important new line, Stephenson recommended a new route, which was approved by the promoters. At the same time, he persuaded Edward Pease, one of the promoters, and Thomas Richardson to join him in establishing a locomotive works at Newcastle.

Railway history was made on 27 September 1825, when Stephenson's *Locomotion* hauled the first train from Stockton to Darlington. As a result, Stephenson was chosen as engineer of the important Liverpool–Manchester railway, the first mainline railway intended to carry passengers. The first step in such a project was to obtain a private Act of Parliament for the route. Opposition could be expected from the owners of the land through which the line would pass, who could usually be bought off. More serious was the opposition of the canal companies and the turnpike trusts, who feared the competition that the railway would bring. When Stephenson gave evidence before the parliamentary committee responsible, counsel for the opposition succeeded in showing that his survey was incompetent and that his proposals were impractical, so that the original Bill for the Liverpool–Manchester line was rejected. However, it was successfully brought back with modifications after a fresh survey conducted by Charles Vignoles the following year.

As well as financial, political and legal obstacles, there were physical problems to be overcome, especially the bogs of Chat Moss. However, just as Jessop succeeded in constructing his aqueduct across the Great Bog of Allen, so Stephenson, advised by his fellow engineers, eventually carried his railway across Chat Moss. In 1827, his son Robert had joined him as manager of the locomotive works and together they designed the famous *Rocket*, which competed in the Rainhill trials to decide the best form of traction to be used on the Liverpool and Manchester line. The *Rocket*, which reached a speed of 30 mph, gained the £500 prize for Stephenson. Tragically, William Huskisson, the parliamentarian who had been most active in promoting the bill for the railway, was killed in an accident on the day when the line was officially opened in 1830.

Thereafter, George Stephenson played a leading part in developing the British railway system, as well as some of the continental ones. To supervise what was eventually to be the Midland railway, he moved to Tapton near Chesterfield. He died there of pleurisy on 12 August 1848. He had lost Elizabeth, his faithful wife through many eventful years, three years before, and took as his third wife his housekeeper, the daughter of a local farmer.

Stephenson was a proud and jealous man, who would acknowledge no peer and brook no contradiction in the field he had made his own. He may have been generous to a fault in the way in which he enabled old associates to share in his success, but woe betide them if they became too successful themselves and claimed a share of the limelight that he enjoyed. Throughout his life, he lost no opportunity of asserting that he fought his way to success entirely single-handed against every kind of opposition from his fellows. He became rich not so much from his engineering work as from successful speculation in South American mining shares. Although he tried to check the railway mania of 1844–6, in which so much money was lost, it was through his connexion with the financier George Hudson that he gave his backing to schemes that had little chance of success.

The first biography of an engineer in any language was *The Life of George Stephenson* (Smiles, 1857). This was largely based on information supplied by Stephenson's son Robert, who worshipped his father. The success of this biography encouraged Smiles to write his three-volume *Lives of the Engineers*, originally published in 1862, three years after the death of the son. The third volume consists of a revised version of the 1857 book, adding '*Robert Stephenson's Narrative of his Father's Inventions, etc.*' as an appendix. Jarvis (1997) points out that there is no mention of Trevithick, and comments that this was just one instance of the way the son credited his father with achievements that he must have known were properly those of other engineers.

Charles Babbage (1791–1871)

The 'irascible genius' Charles Babbage was born in Southwark, on the opposite bank of the Thames to the City of London, on 26 December 1791, one of two surviving children of the banker Benjamin Babbage and his wife Betty Plumleigh, née Teape, both members of well-known Devonshire families. His father, an enlightened man, made no attempt to influence the boy's determination to become a scientist rather than a banker. His mother was equally enlightened. Physically he was a delicate child but mentally he was advanced in development, showing exceptional intellectual curiosity. He spent his early childhood in London, where the family lived over the father's bank, but soon after his 11th year, following a severe illness, he was sent first to stay with a clergyman in the country near Exeter, in the hope this would be good for his health, and then to his father's old school, King Edward VI Grammar School at Totnes, where he received a classical

education. As adolescence approached, his health improved and he attended a small private academy, headed by another clergyman, north of London. Although it was not a good school, there was an excellent library, which he took advantage of. Next, he was sent to tutors to prepare for university entrance. In October 1810, when he was nearly 19 years old, he was admitted to Trinity College Cambridge, but finding the competition from the other Trinity undergraduates too strong for him he transferred to Peterhouse. He was good at mathematics and came top of the list of Peterhouse undergraduates in the Tripos examination.

At Cambridge, much time and effort were devoted to learning how to deal with the particular kind of mathematical problems set in the Tripos examination, which were only rarely related to any physical questions not considered in Newton's *Principia*. Some of the undergraduates, including Babbage, were dissatisfied with the tuition offered, which consisted of the rote learning of part of *Principia* and certain other British texts. They formed the Analytical Society, to reform the Cambridge syllabus and, later on, the Royal Society, which had degenerated into too much of an exclusive social club.

In 1814, the year before he took his degree, Babbage married the charming young Georgiana, daughter of William Whitmore, member of an important Shropshire family. She was 22, one year younger than he was. Obsessed with scientific work, which he hoped would help him secure a suitable academic post; he more or less ignored her. They made their home in Marylebone, where she bore him eight children, of whom only three sons survived to adulthood. Meanwhile Babbage was dependent on his father, who somewhat disapproved of his marriage. When suitable posts became vacant he applied, armed with an impressive list of recommendations, but was unsuccessful. He was asked to organize and later manage a life-insurance society. Although he wrote a book on life expectancy called *A Comparative View of Various Institutions for the Assurance of Lives*, which was quite successful at home and abroad, he declined the opportunity. He was elected a Fellow of the Royal Society in 1816 and contributed mathematical papers to various learned journals. On a visit to Paris with his friend John Herschel, he met Laplace, Fourier and other French scientists.

Babbage was a founding member of the Cambridge Philosophical Society, successor to the Analytical Society, and the Astronomical Society, later to become the Royal Astronomical Society. It was while calculating some tables for the latter that he resolved to make what he called the Difference Engine, which would do the tedious work without human intervention, printing out the result automatically. It was well known that there were a huge numbers of errors in virtually all sets of tables then in use. He was just over 30 when he announced this in a paper to the Society at a meeting in 1822, and informed the President of the Royal Society, Sir Humphry Davy, of his intention. When he applied to the government for funds to build the machine, the government asked the Royal Society for a report, which was favourable. The Chancellor of the Exchequer told him that although in general the government were unwilling to make grants of money for any inventions, however meritorious, his case was exceptional, because mathematical tables were peculiarly valuable for nautical purposes. Soon after this interview, the Royal Society was informed that the Treasury had granted Babbage £1,500 to enable him to bring his invention to perfection in the manner recommended. Further grants followed, until Babbage had received public funds totalling £6,000.

He now embarked on the construction of his machine, expecting that it would take two or, at most, three years to complete. He realized that special tools would be required: to make them he secured the services of Joseph Clement, one of the foremost tool makers of the day. In 1827, overwhelmed by a series of personal tragedies, Babbage suffered a nervous breakdown. His medical advisers urged travel abroad for six months, and he took with him a workman named Richard Wright, who was to prove a faithful friend. It was arranged that Georgiana, again pregnant, would live with her sister. She died after giving birth to another boy. His father had died leaving him with a life interest in his valuable estate. Although he could not touch the capital, Charles was comfortably off.

He set off for Venice, via Holland and Germany, meeting many interesting people on the way. He went on to Rome, where he was very surprised to read in a newspaper that he had been elected Lucasian professor of mathematics at Cambridge. After some hesitation he accepted the chair, which he held for ten years without ever giving a lecture. He was elected to the Royal Academy of Naples, climbed Vesuvius, and saw the sights. On the way back to England he went to Berlin, where he was welcomed by the polymath Alexander von Humboldt and introduced to other German scientists who were gathered in the Prussian capital for a conference.

On his return, he found trouble. Word had been spread around that the *Difference Engine* project, on which public funds had been expended, had been aborted. In fact, its construction was making steady progress and he judged it was three-fifths complete. Having spent much of his own money on the project he applied for another grant from public funds. It was agreed to advance £3,000 on top of the previous £6,000, on condition that the machine, when complete, would belong to the government. Also professional engineers would be appointed to examine the accounts, which was acceptable to Babbage although it introduced delays in the process of payment. He was offered a knighthood, but rather tactlessly turned it down.

Babbage was hoping to be appointed junior secretary of the Royal Society, his friend Herschel being the senior. He was not, probably because he quarrelled with Davy, the president, who claimed the right of nomination. Feeling cheated, he wrote a polemical work, which was published in 1830 under the title *Reflections on the Decline of Science in England and on Some of its Causes* (Babbage, 2009a). He began by criticizing the system of education, which he correctly argued was in need of radical reform, but then launched an attack on the Royal Society. This made him dangerous enemies, especially Airy, the Astronomer Royal, who was already unfriendly towards him. In 1831, he compiled 21 volumes of logarithms of the first 108,000 integers; great pains were taken to achieve clarity in this by varying the typeface and colour of the paper. By this time, Babbage had changed the design of the Difference Engine so much that it was essentially a new machine.

Unfortunately, his employee, Joseph Clement, was becoming dissatisfied. He was a first-rate craftsman, whose work was of the highest standard, and he charged accordingly. The engine was at Clement's workshop in Lambeth; Babbage wanted it moved to his house in Manchester Square, several miles away, where he could see what was happening, but naturally this did not suit Clement. Also, the Treasury was slow at paying Clement's bills, because they now had to be certified by the professional engineers, as agreed. Clement ceased work and dismissed his team of workmen. He also retained the tools he had made to Babbage's orders, and the drawings that went with them, which in law he was entitled to do. From 1854 onwards, all work on the *Difference Engine* was suspended, and in fact it was never to be resumed. That this, like many others of his projects, was never brought to a state of completion was in part owing to his own character, in part to lack of financial backing, and in part to the inadequacies of existing precision technology.

Babbage had been involved in politics for some time and now stood for parliament as a liberal at the next election in 1832. He was bottom of the poll. The same year, he published what is regarded as a pioneer work on operational research, entitled *On the Economy of Machinery and Manufactures* (Babbage, 2007). To gather material for this, he had visited numerous factories and workshops on the continent as well as in England. He became the leading advocate of the systematic application of science to industry and commerce. He attended the early meetings of the British Association, of which he was a trustee. Since there was no room for statistics at the meetings, he founded the Statistical Society of London, and became its Chairman. In 1834, his only daughter Georgiana died in her late teens. He was left with the three boys, but seems to have taken no particular interest in their upbringing.

Another of his multifarious interests was the new railway system. Trains on these early railways were unsafe and uncomfortable. He made several proposals for improving the safety of railway travel; signalling systems were rudimentary, early locomotives had inefficient brakes and there were frequent accidents. Like Cayley, Babbage suggested some improvements to make rail travel safer and more comfortable. For this purpose, he constructed a dynamometer car, to be attached to a train, which would record the speed, vibration and other characteristics of a journey.

Babbage's main claim to fame is based not on the *Difference Engine*, one of a number of similar machines that were produced in that period, but on his plan for constructing another machine, based on an entirely different principle from that of the *Difference Engine*. As in a modern electronic digital computer, but mechanically, the operation was to be programmed by means of punched cards, rather in the way that they were used by Joseph-Marie Jacquard in the loom bearing his name. This machine, which Babbage called the *Analytical Engine*, can properly be described as a computer, rather than just a calculating machine. While the *Difference Engine* was designed to work straight through a computational problem, the new machine was designed to make calculations, store the results, analyze what to do next, and then return to complete the project. Thus, it could carry out complicated operations of arithmetic, which the *Difference Engine* could not.

When Babbage informed members of the government of his plan, without asking for funds to carry it out, they were unhelpful, especially when Airy gave his opinion that the concept was quite worthless. In 1838, feeling exasperated, Babbage went to a meeting of 'philosophers' in Turin, where his idea was greeted with enthusiasm. Encouraged by this, he began constructing more detailed plans for the *Analytical Engine*. He decided to try and construct the engine at his own expense, employing draughtsmen and workmen, maintaining in his own house the investigations and experiments necessary to accomplish his supreme work of genius.

Only parts of it were built: the complete machine would have been the size of a railway locomotive, and would have required steam power to operate it.

In his prime, a man of immense personal charm and boundless vitality, Babbage was very sociable and became a leading figure in English society. His regular Saturday soirées were attended by some of the most interesting people of the time, both English and foreign. One of them was Ada, Countess of Lovelace, who was the only legitimate child of the poet Lord Byron and, like him, suffered from manic depression. At the age of 19, she married a good-natured and scholarly man of means who in 1838 was created Earl of Lovelace. Ada had passionate, quite immoderate, intellectual interests: she loved mathematics and music; also, 'I am afraid that when a machine, or a lecture, or anything of the kind comes in my way, I have no regard for time, space and any other obstacles.' She was impressed by the *Analytical Engine* and saw the great beauty of the invention; her interest in it led to a lifelong friendship with Babbage. With her friend Mary Somerville, known as the Queen of Science, she often went to Babbage's house in Marylebone, while he was a frequent guest in the Lovelace household.

In 1842, a treatise on Babbage's invention had been published in French. Ada had developed enough confidence in her mathematical abilities to undertake an English translation, with her own commentary, which expanded the treatise to three times its original length. In a famous and influential metaphor, she wrote that the *Analytical Engine* weaves algebraical patterns just as the Jacquard loom weaves flowers and leaves. Although Babbage advised her on substantive matters, Ada was very proprietary about the work, chastising him when he suggested changes. Her husband also helped with the work by copying and making himself useful in other ways.

Ada was very proud of her book, which appeared in 1843. She praised her own masterly style and its superiority to that of the original memoir. Her work on the project, especially her programme for using it to compute Bernoulli numbers, has earned her the designation of being the first computer programmer. She asked penetrating questions as to how the engine might be applied, and conjectured that, 'If it could understand the relations of pitched sounds and the science of harmony, the engine might compose elaborate and scientific pieces of music of any degree of complexity and extent.' She realized the potential of the *Analytical Engine* much better than Babbage himself, although it would be over a hundred years before her ideas became a reality.

Ada, like her father, had a mercurial temperament that swung precipitously from the ecstatic and grandiose to the melancholic. Like him, she enjoyed gambling and ignored the financial chaos that this created. In 1850, she took up betting on horse-racing, using a mathematical system she had devised with Babbage, and was soon embroiled in legal and financial difficulties as a result.

Babbage became increasingly bitter at the lack of interest in the *Analytical Engine* by members of the government, particularly the prime minister, Sir Robert Peel. Babbage wrote more books, one of which argues the case against hereditary peerages and recommends life peerages instead. In 1823, he had been awarded the Gold Medal of the Royal Astronomical Society, but otherwise he collected surprisingly few honours, mainly because he declined those he was offered. For example, Peel offered Babbage a baronetcy, which he declined. Among his last works was an entertaining autobiography *Passages from the Life of a Philosopher* (Babbage, 2009b). It was felt he had lost all sense of proportion when he promoted a parliamentary bill restricting the activities of organ-grinders, who disturbed his work. He died on 18 October 1871, within three months of his 80th birthday.

Apart from the design of the *Analytical Engine*, which possessed many of the features of the modern computer, Babbage also pioneered machine tools, lighthouses, codes, stage lighting, an ophthalmoscope, postal services, operations research and much else. The American scientist Joseph Henry, whose profile follows later, gave a picture of Babbage's minor contributions:

> Hundreds of mechanical appliances in the factories and workshops of Europe and America, scores of ingenious expedients in mining and architecture, the construction of bridges and boring of tunnels, and a world of tools by which labour is benefited and the arts improved – all the outflowings of a mind so rich that its very waste became valuable to utilize – came from Charles Babbage. He more, perhaps than any man who ever lived, narrowed the chasm separating science and practical mechanics.

Charles Blacker Vignoles (1793–1875)

Another engineer with French antecedents was Charles Vignoles. He was born on 31 May 1793 at Woodbrook, county Wexford, the only child of Charles Henry Vignoles, a descendant of a Huguenot military family

serving in Ireland, and Camilla, née Hutton. His father was a captain in the 43rd or Monmouthshire Regiment of Foot, which was sent out to the West Indies; he was wounded and taken prisoner at the storming of Pointe-à-Pitre in Guadeloupe in 1794. Shortly afterwards, he and Camilla died of yellow fever.

The orphan Charles Vignoles was brought to England by an uncle and raised by his grandfather. When 18 months old, Vignoles was gazetted as an ensign on half pay in his father's regiment, a common method of compensating deceased officer's families. He was educated at the Royal Military Academy at Woolwich, where his grandfather taught, but since his grandfather favoured a legal career he was articled as a proctor in Doctor's Commons for seven years. How long he remained there is uncertain but in 1843 he had a dispute with his benefactor, which was never healed.

In 1813, Vignoles was transferred to the York Chasseurs and soon afterwards was at Sandhurst as a private pupil of Thomas Leybourn, one of the college lecturers. Leybourn was guardian of Mary Griffiths, eldest daughter of a Welsh gentleman farmer and Charles and Mary became secretly engaged. Nominated by the duke of Kent, Charles was commissioned in the 1st or Royal Scots Regiment in 1814. He saw action at Bergen op Zoom in March and then in Canada for eight months. In 1815, he was made lieutenant and posted to Scotland, where he became aide-de-camp to General Sir Thomas Brisbane at Valenciennes. During his stay there he compiled tables for comparing French and English weights and measures at the request of the Duke of Wellington.

Reduced to half pay in 1816 and without private means, Vignoles had to earn a living somehow. After returning to England he married Mary Griffiths at Alverstoke, Hampshire, in 1817 but almost immediately set sail for America, intending to serve under Simón Bolívar in the wars of independence. However, it turned out that Bolívar did not need his services and by the end of the year he was at Charleston, South Carolina, as assistant to the State's civil engineer. In 1821, he became city surveyor of St Augustine, Florida, and in 1823 published a map of Florida. Severe financial problems and news of his grandfather's death persuaded him to return to England in the same year.

In England, Vignoles worked as a surveyor and wrote articles for the *Encyclopaedia Metropolitana*, before becoming assistant to James Walker, engineer of the London Commercial Docks. By mid-1824, he had his own office in Hatton Garden, with three pupil assistants. In 1825, he was hired by the Rennies, father and son, to survey a railway between London and Brighton, and a new Liverpool and Manchester line, after Parliament had rejected George Stephenson's original scheme. Vignoles moved north and made Liverpool his base for the next 15 years. His legal training made him a good parliamentary witness, a factor in the success of the second Liverpool and Manchester Railway bill, and it brought much similar work subsequently. He was employed as Stephenson's assistant but a clash of personalities made him resign in January 1827 after a disagreement over measurements for the Edge Hill tunnel. Marc Brunel offered Vignoles a post as resident engineer on the Thames Tunnel but withdrew the offer when he realized that he could appoint his own son instead.

Vignoles subsequently spent a year on the Isle of Man, surveying government property. Brunel recommended him to straighten the northern section of the Oxford Canal, but they fell out after Vignoles criticized Brunel's work on the breached Thames Tunnel. In 1829, Vignoles, in collaboration with John Braithwaite and John Ericsson, competed in the Rainhill trials with the lightweight locomotive *Novelty*. Although this was the favourite, it broke down and was beaten by the *Rocket*, as we have seen. In 1830, he and Ericsson patented a method of ascending steep inclines on railways. Vignoles also advocated a flat-bottomed rail, which would bear directly on sleepers without any chair, but despite occasional trials the rail was never adopted in Britain. It was adopted on the continent, and is named after him in Germany and France.

During the 1830s, Vignoles surveyed and constructed numerous English railways, especially in Lancashire, and also worked in Ireland, France and Germany. Between 1832 and 1834, he worked on Ireland's first railway, the Dublin and Kingstown line, and between 1836 and 1838 he was engineer to the Royal Commission on railways in Ireland. At the end of the decade, financial problems almost ruined him. In 1835, he had surveyed the proposed Sheffield, Ashton-under-Lyne and Manchester Railway and became the resident engineer. When problems arose with raising the necessary finance, Vignoles, with the blessing of the board of directors, bought many of the depreciated shares in the names of friends and relatives, on the understanding that they were trustees for himself and that no calls for further investment would be made on them. This manoeuvre enabled work to commence, but the directors later insisted after all that calls had to be met, leaving Vignoles with a bill for £14,000 he could not meet. He fought the decision but had to resign. Although he eventually paid off these debts, the next three years were difficult. He became professor of engineering at University College, London, advocated and built atmospheric railways, and reported on railways in the German kingdom of Württemberg.

Although Vignoles had achieved a great deal, he still had no spectacular engineering achievement to his credit until in 1846 he was employed to construct the new bridge across the Dnieper river at Kiev. He moved to the site and lived there until the bridge was completed in 1853. During these years, Vignoles paid several visits to England and in 1849 he married Elizabeth Hodge at St Martin-in-the-Fields, his first wife having died in 1834. After the Kiev Bridge, Vignoles was involved in some English projects but his main work was abroad, most notably the Frankfurt, Wiesbaden and Cologne Railway in Germany, the Western Railway in Switzerland, the Bahia and San Francisco Railway in Brazil, and the Tudela and Bilbao railway in Spain. In 1863, he retired, and in 1867 acquired a house in Hythe, near Southampton, where he became a country gentleman and Justice of the

Peace. He still visited London frequently, actively participating in several scientific societies. On his return from one such visit, he suffered a stroke and died four days later at his home on 17 November 1875. From his first marriage there were seven children, five of whom reached adulthood. Three of his sons entered the engineering profession; another became a minister of the Church of England and wrote a biography of his father.

Vignoles joined the Institution of Civil Engineers in 1867 and became president in 1869. He became a fellow of the Royal Astronomical Society in 1829, a fellow of the Royal Society in 1855, was a founder member of the Photographic Society of London, served in 1855 as a member of the Royal Commission on the Ordnance Survey and was connected with the Royal Irish Academy and the Royal Institution.

Sadi Carnot (1796–1832)

The life of Lazare Carnot was profiled in a previous chapter. His eldest son Nicolas Léonard Sadi was born in Paris on 1 June 1796, not long after two elder sons, also named Sadi, had died in infancy. The birth took place in the rooms of the Petit Luxembourg, which the family occupied, and the first

year of Sadi's life was spent there. In September 1797, however, the family was split up. Following the coup d'état in that year, Lazare, as we know, went off alone into exile, while Sadi went to live with his mother, Marie Dupont, at her family home in Saint-Omer. Even when his father returned to Paris in January 1800, Sadi can have seen little of him, and it was only after Lazare's resignation from the Ministry of War in September 1800 that normal family life became possible. Lazare was an attentive father and the years from 1801, when his second son Hippolyte was born, until the death of his wife in 1813 were obviously happy.

Sadi benefited greatly from his father's leisure: in fact he was taught by him until he was nearly 16. Although the education he received reflected Lazare's literary interests, Sadi's aptitude for science and mathematics was encouraged and in November 1812 he entered the Ecole Polytechnique after some special preparation by a noted teacher of mathematics at the Lycée Charlemagne. In the fiercely competitive entrance examination, Sadi was ranked 24th out of 184 successful candidates, a high position for someone of his age; in his promotion there were only two others who were younger.

Carnot was a good and diligent student but not outstanding. A contemporary described him as extremely gentle: 'He behaves well and is a little shy, a very rare quality in the youth of today.' Confined to their barracks in the old Collège de Navarre on the Montagne Sainte-Geneviève they were closely supervised in all they did. Carnot's second year was seriously disrupted by the war. On behalf of his fellow students, Sadi addressed a letter to the emperor asking permission to join in the fight against the invading allies. When this was granted, the polytechniciens gave themselves almost entirely to military training, and when the time came for the final defence of the capital late in March 1814, they faced the enemy on the road to Vincennes with a courage that has become legendary. However, Paris eventually fell to the allies and Bonaparte was forced to abdicate. Carnot visited his father in exile, no doubt to inform him of the situation and receive his advice on what to do.

Along with some of the best graduates of his year, in 1815 Carnot entered the Ecole d'Applications de l'Artillerie et du Génie at Metz, where he studied until he was commissioned as a second lieutenant of the Corps Royal du Génie in 1817. His career as an officer was undistinguished. He was sensitive and shy, quite unsuited to the boisterous life of an army garrison. He contrived to spend a good deal of his time in Paris, where he was transferred to the newly formed administrative Corps d'Etat-Major, with the rank of lieutenant. From 1819, he was able, apart from a short annual military exercise, to resume residence in the Marais, where he lived for almost the whole of the rest of his life. He attended lectures at the Sorbonne and at the Collège de France. He was a frequent visitor to the Conservatoire Nationel des Arts et Métiers, where courses of lectures on mechanics, industrial chemistry and industrial economy were provided. He made frequent visits to factories and workshops to observe industrial processes.

In 1821, Carnot spent a few weeks with his father in Magdeburg, after which he began composing his masterpiece and unique published work, the *Réflexions sur la puissance motrice du feu*, in which he tried to discover the general operating principles of steam engines and other heat engine machines that supply work output from heat input. Carnot was motivated by the realization that the enormous industrial importance of the steam engine and the great improvements that had been effected in its design by a succession of British engineers, was matched by almost total ignorance of the theory behind its working. The *Réflexions* were inspired by his father's attempt to provide such a theory. Soon after its publication it was presented to the relevant section of the Institut de France and a favourable report was obtained, but it was not a success. Physicists were dismissive, because of the lack of mathematical analysis, and so were engineers, because of the lack of practical applications.

In June 1832, Carnot fell ill, apparently with scarlet fever. He recovered from this but after a relapse he was moved to the exclusive private clinic of an alienist because he was also experiencing manias. In his weakened state, on 24 August, he succumbed to the cholera which was then raging. As was customary his personal effects, including his scientific papers, were burnt, thus causing great frustration to historians of science. Enough survived to show that much of value must have been lost.

Carnot's work might have been entirely overlooked had not William Thompson, the future Lord Kelvin, in Britain, and Rudolf Clausius, in Germany, discovered it and been inspired to formulate, independently, the second law of thermodynamics. Thomson wrote, 'Nothing in the whole range of natural philosophy is the more remarkable than the establishment of general laws by abstract reasoning.'

4 From Henry to Bazalgette

Joseph Henry (1797–1878)

There are some experimental physicists who might also be classified as engineers. I shall profile a few of these, beginning with the American Joseph Henry. He was born in Albany, the state capital of New York, on 17 December 1797. His father, William Henry, was a sometime day labourer from Argyle, distantly related to the earls of Stirling, while his mother Ann, née Alexander, was a miller's daughter. William Henry died young, and it was chiefly his widow Ann who brought up her son. She was a small woman with rather delicate features who lived to an advanced age. A devout and strict member of the Scottish Presbyterian church, she passed its Calvinist principles on to her son. Before he had turned six she sent him to nearby Galway to live with her stepmother and her twin brother John.

After three years of elementary school, Joseph took a job in a general store where the shopkeeper, an educated man, encouraged him to continue with his education after work. When the boy was approaching 14, he returned home to Albany, where he was apprenticed to a watchmaker and silversmith. After the business had failed, he was released from his apprenticeship but not before he had acquired some practical skills, which were to be useful to him later.

Although Albany was not a large place in those days, it boasted an unusually good theatre, which for a year occupied most of the young man's leisure hours. He belonged to an amateur group for which he acted, produced and wrote plays. The experience may have helped him in later years when he needed to be an effective public speaker. During this period, Joseph Henry is described as being remarkable for his good looks, delicate complexion, slim figure and vivacious nature. His lively temperament made him a general favourite. The town had an excellent school, almost a college, called the Albany Academy, where Joseph Henry continued his education by attending night classes in geometry and mechanics, followed by calculus and chemistry, while supporting himself by working as a private tutor and doing a little teaching.

Albany also possessed a scientific society, the Albany Institute, with a miscellaneous library of some 300 volumes. Henry obtained the post of librarian and gave a few lectures to the members, complete with experimental demonstrations. Meanwhile, he was studying the classic *Mécanique Céleste* of Laplace. The versatile young man also led a grading party for a new highway running from the Hudson river to Lake Erie. His success in this work brought him offers of similar positions elsewhere but he turned them down in favour of the post of professor of mathematics and natural philosophy at Albany Academy in 1826.

A maternal uncle of Henry's, a successful businessman in Schenectady, had died and his widow brought her family to live in Albany. Her son Stephen was a delicate youth, who at the age of 18 had graduated from Union College with high honours in mathematics and astronomy. Since they had similar interests, the two cousins saw much of each other; soon Stephen was also on the faculty of the academy and later he would follow Joseph to Princeton. Meanwhile, Joseph had been courting Stephen's sister Harriet and they were married by the end of 1829. Modest, retiring and understanding, Harriet furnished the home into which her husband would retreat from the pressure of the outside world. She seems to have been an ideal wife for him. They had six children, of whom two died in infancy and the only son died on the threshold of manhood. The three survivors, Helen, Mary and Caroline, were to brighten their parents' declining years.

In his 36th year, Henry was appointed to the chair of natural philosophy at the College of New Jersey, which would become Princeton University many years later. When he started work there in 1832, the college was at a low ebb, with only about 70 students, but the situation was beginning to improve and later he was to say that the 14 years he spent at Princeton were the happiest of his life.

One advantage of Princeton was that it brought him into contact with a wider circle of men engaged in scientific research. No longer was he obliged to work in an intellectual vacuum. For example, a short journey on the newly opened railroad brought him to Philadelphia, seat of another university, and home of several learned societies. One of them was the American Philosophical Society, to which he was elected in 1835. It was to members of this society that he communicated the results of his Princeton researches. Unfortunately, the society was rather slow in publishing the papers presented to it, so Henry was continually at a disadvantage with his rival Faraday, who took no interest in what was being done in America. Faraday was much more favourably situated at the Royal Institution in London, living in the world centre of intellectual activity, with as much time as he wanted for research, in constant communication with some of the most brilliant minds of the day.

In 1836, in recognition of his work during his first four years at Princeton, the trustees of the college granted Henry a year's leave of absence with full salary. It was a good time to escape from the United States, for the nation was just entering an economic depression. England was, in many respects, a different country from the one which the American colonists had rebelled against. Reform was in the air and the spirit of industry had taken a firm grip on the people. Although London was his destination, Henry began his journey by first going to Washington, to collect some letters of introduction; he was not impressed by the federal capital, where he was destined to spend so much of his life.

Once in London, Henry lost no time before visiting the Royal Society and the Royal Institution. Unfortunately, it was the Easter vacation and Faraday had just left town, but they met before long and Henry heard him lecture. After two useful and enjoyable months, he moved on to Paris for two weeks. After brushing up his French, he met a few of the French men of science and attended a meeting of the Paris Academy, but otherwise he seems to have made little real contact with the leading French scientists. Finally, after a short visit to the Netherlands, he returned to London to stay with his former student Henry James, father of the psychologist William and the well-known novelist.

Both Joseph and Harriet being of Scottish descent, like so many New Englanders, they took the packet to Edinburgh and called on many of the notables there as well as elsewhere in Scotland. His tour ended in Liverpool, where a meeting of the British Association for the Advancement of Science was in progress. He came back to Princeton in excellent form, ready to resume his own investigations into electromagnetism. For ingenuity, completeness and novelty, these researches serve as a model for the experimental physicist.

In his 14 years at Princeton, Henry had found congenial companions and duties well-suited to his powers. He had been valued and honoured by members of the faculty, while the students held him in reverence. Although far from exhausted scientifically, rather at the height of his powers, in his 49th year, he chose to withdraw from active scientific work and became an organizer and administrator. The occasion that brought about the change was his decision to accept the post of secretary of the nascent Smithsonian institution, at the end of 1845. Most of his friends advised against his moving to Washington, pointing out his lack of relevant experience, and Henry himself felt no great enthusiasm. He was promised that Princeton would welcome him back if he decided that he had made a mistake. He turned down the offer of a chair at the University of Pennsylvania, probably the most desirable scientific position in the country. Henry hoped that he would be able to return to experimental work once the Smithsonian got under way but that was never a real possibility. His annual salary for the arduous work of establishing the new institution was a mere $3,000, but in the 32 years he served as director he would never accept an increase.

Before long, the many-turreted red sandstone building in vaguely Romanesque style known as 'The Castle' was under construction on a site in Washington. Henry began to organize a series of publications and a system of exchanges with other institutions that enabled the library to develop. The subjects in which he took an interest were astonishing in their variety. For some, such as meteorology, he was enough of an authority himself but for others, such as zoology, he consulted a network of outside experts. Increasingly, the federal government used the Smithsonian as a general-purpose institution, for example, relying on Henry to organize the land surveys and expeditions that were so necessary for the exploration of the American continent. Advice was freely given when telegraph lines were to be laid or railroads planned: this brought out the engineer in Henry.

The family, consisting of Joseph, his wife and their three daughters, resided in the Castle itself, where they entertained liberally. The building was badly damaged by fire in 1865 and much valuable material destroyed, including Henry's voluminous correspondence and research notes. After the end of the American Civil War, he made another visit to Europe with his daughter Mary. He continued in office until the age of 80 and was still at work at the end of 1877, when he found one morning that his right hand was paralyzed. Nephritis, in those days incurable, was diagnosed, and he knew that he did not have long to live. He died in his sleep on 13 May 1878 and was buried in the Oak Hill Cemetery in Georgetown, Washington.

Henry was the first to elucidate the principles of the design of electromagnets and constructed the largest then known, which could lift 300 kilograms. He was the first to employ them in a successful telegraph system, and the first to construct a reciprocating machine driven by direct electric current. These inventions were fundamental to the development of the electric telegraph and electrical machinery. On principle, he never patented any of his inventions, which when exploited commercially made others rich but not the inventor. As he explained:

> My life has been principally devoted to science and my investigations in different branches of physics have given me some reputation in the line of original discovery. I have sought, however, no patent for inventions and solicited no remuneration for my labours, but have freely given their results to the world, expecting only in return to enjoy the consciousness of having added by my inventions to the sum of human knowledge. The only reward I ever expected was the consciousness of advancing science, the pleasure of discovering new truths, and the scientific reputation to which these labours would entitle me.

The value of his scientific work was only properly appreciated after his death. Its significance lies not so much in the material he added to science as in the new territories he opened to the view of those who followed in his footsteps. In recognition of his achievements, the unit of inductance is named after him.

John Ericsson (1803–1889)

John Ericsson has already been mentioned as one of the unsuccessful competitors in the Rainhill trials. The story of his life begins in Sweden, where he was born on 31 July 1803, at Färnebo in Värmland, in the centre of the country. His forefathers had been miners; in particular his father was a mining engineer, who was already in charge of a mine at the age of 21. Ericsson inherited from him an enthusiasm for technology and from his tall, slim and good-looking mother he inherited his noble outlook and strong character. He was the youngest of three children; he had an elder brother, Niels, who also became an engineer.

Sweden lost Finland in the Russian war and trade suffered severely. In the economic decline their father lost his job in mining and although he obtained another as inspector during the building of the Göta canal, he died when his son John was 15. Already, the young man was studying the machinery used in mining. Count von Platen, a friend of his late father's, helped him get started in life. After spending two years training with the Naval Engineers, Ericsson was commissioned in the Swedish army. Tall, well-set, healthy and strong, he wanted to serve his country, which was frequently at war. In the artillery, he learned how to handle large guns,

knowledge that was of great value to him professionally later. He also studied chemistry and mathematics, including geometry and surveying.

He resigned from the army, with the rank of captain. Like many others at the time, he thought the steam engine could be improved. He designed a hot-air engine, with no boiler, and took it to England in 1826, but no-one was interested in it because its consumption of coal was enormous. Also, a Scottish parson named Stirling had previously obtained a patent for a similar engine. However, the machine manufacturer John Braithwaite took him into junior partnership; the firm called itself Braithwaite and Ericsson. Here Ericsson learned much about machine design. He attempted unsuccessfully to use compressed air for power transmission. He built boilers and refrigerators, and designed a surface condenser, which he successfully applied to marine engines. He built high-pressure boilers and, in designing ships, tried to get the engines installed below the waterline, to protect them from attack in case of war. Finally, he designed and constructed a steam engine, which he built in London in 1828. Several were built; one of them was bought for the Royal Palace in Berlin. However, it was not a commercial success. Ericsson continued to try to improve the steam engine, for example, by constructing a rotary engine, which rotate around a fixed crankshaft. He took out three or four patents a year over a long period. Ericsson's marine engines were particularly successful. He also designed a revolving screw to propel ships, in place of the paddle wheels, which were in common use. Several others had tried this out but it was believed that such ships could never be properly steered, until one driven by Ericsson's propeller had successfully crossed the Atlantic.

Ericsson had greater success when he emigrated to America. A number of his screw-propelled steamers were in service on the Great Lakes, and his great screw-propelled frigate *Princeton* became famous. In 1852, he completed another great ship, named after himself, this time powered by one of his hot-air engines. Unfortunately, it consumed at least as much coal as a steam engine, broke down on its maiden voyage and sank with the loss of all its crew. However, the hot-air engine's use to power small vessels was more successful; over 3,000 were manufactured.

War has always stimulated progress in engineering. When the American Civil War started in 1861, the Confederate states were in command of almost the whole of the United States Navy. In a short space of time, 500 men-of-war were constructed for the Navy of the Union, but the Confederacy still had the upper hand at sea. Ericsson got an order from the Union Navy for a turreted ironclad called the *Monitor*, which was so successful he was asked to make six more; others were ordered in England and Sweden. He continued to design warships, torpedoes and other naval weapons, although he never lost interest in his hot-air engines. In the closing years of his life, he was exploring the possibilities of solar energy, gravitation and tidal forces as sources of power.

Ericsson was a loner and a workaholic. He had married in England in 1836 but his wife soon returned home from America in the hope that he would join her. He never did, although he supported her all her life. Even in his 80s he enjoyed good health. He died on 8 March 1889. When his coffin was brought to Sweden, 64 years after he left, he was buried in his homeland with full military honours.

Robert Stephenson (1803–1859)

It is not difficult to find instances of remarkable engineers whose sons were also remarkable engineers. The Brunels are one example, the Carnots another, and the Stephensons a third. Robert, the only child of George

Stephenson's second marriage to survive, was born at Willington Quay on 16 October 1803. As we have seen, his father, who had not yet become famous, moved to Killingworth the following year, where Robert attended the village school, then went to an academy at Newcastle, before being apprenticed at the Killingworth colliery where his father worked. His career really starts when he assisted his father in the survey of the route for the historic Stockton and Darlington railway. After spending a session at Edinburgh University, he settled down in Newcastle to manage the locomotive factory that his father had founded. During this period, he designed and built the locomotive *Active*, which was later renamed *Locomotion*. Against his father's wishes, in 1823 he went out to Colombia on a three-year contract to superintend the working of machinery in some gold and silver mines. When he returned to Newcastle in 1827, the famous *Rocket*, which won the Rainhill trials, was built at the Newcastle works under his direction.

Obtaining parliamentary permission for the London–Birmingham railway involved him in much work giving evidence to parliamentary committees in 1832. His appointment as engineer-in-chief the following year marked his emergence as an engineer in his own right. He moved to London the better to cope with the problems posed by the 112-mile line, from tunnels at Primrose Hill and Watford, to fighting the canal company at Wolverton, and an epic fight against subterranean quicksands at Kilsby. The line was opened in 1838. Stephenson also took an active part in the 'battle of the gauges' and in the struggle between rival advocates of the atmospheric and locomotive systems. Up to 1840, the works sent locomotives to France, Germany, Italy, Belgium and the United States; for the rest of his life, Stephenson was involved in the construction of railways abroad as well as at home.

During a four-year period, ending in 1847, promoters presented hundreds of schemes to Parliament to build new routes. Almost 10,000 miles of new lines received approval but a third of this mileage never materialized. Most of those built eventually proved viable: one route in particular, the Great Northern route linking London and the north of England, was especially valuable. The new network greatly improved the economics of many industries, although Stephenson reckoned that a network just as productive could have been built for a third less than the actual cost. During the railway mania, less than a tenth of the planned mileage was ever constructed and speculators often lost their investment.

Although Robert Stephenson was one of the principal figures behind the expansion of Britain's railways, he is chiefly famous as a designer of bridges, especially railway bridges, which had to carry much heavier loads than highway bridges. His first, across the river Dee at Chester, collapsed with a train on it but soon after he built the splendid High Level Bridge over the river Tyne at Newcastle and the Royal Border Bridge at Berwick. His most famous railway bridge was the Britannia tubular bridge, across the Menai Straits, using wrought iron, with two 459-foot spans. He also built the similar railway bridge at Conway, the great Victoria Bridge over the St. Lawrence river at Montreal, for many years the longest bridge in the world, and many others.

In the last decade of his life, Stephenson moved away from engineering and increasingly became a man of affairs. Known as a man of sound judgement and unimpeachable integrity, he was appointed to numerous parliamentary committees. A Tory of the old-fashioned protectionist school he served as member for Whitby from 1847. In Parliament, when he spoke on engineering matters, his opinion was often decisive; for example, he was opposed to the construction of the Suez Canal. As a member of the Royal Commission responsible for the Great Exhibition of 1851, he was in favour of Paxton's design for the building, which became called the Crystal Palace. He served as president of the Institutions of Electrical Engineers and Civil Engineers. Oxford University awarded him the honorary degree of DCL. He also received many foreign honours.

Stephenson married Fanny Sanderson, the daughter of a London merchant, in 1829. They had no children. After a few years in Newcastle, they moved to London in 1834, where their home was on Haverstock Hill, Hampstead. After his wife died in 1842, he moved to Cambridge Square, just north of the Bayswater Road, and six years later, having become sufficiently wealthy, to a much grander house in Gloucester Square, where he entertained lavishly and began to collect works of art, especially portraits of engineers or paintings with some connection to his work. He was elected a fellow of the Royal Society in 1849. His main recreation was sailing; he owned a 188-ton ironclad yacht, the *Titania*.

Robert Stephenson suffered from poor health for much of his life. He died on 12 October 1859 at home in London, after falling ill while cruising in Norwegian waters aboard the *Titania*. Like Telford and Watt, he was buried in Westminster Abbey. Thousands thronged the streets along the route of the cortège as it passed on its way to the national pantheon. The Queen gave unprecedented permission for it to pass through Hyde Park.

The Times newspaper, in its eulogy, described Stephenson as:

> ... the eminent engineer, distinguished not more for his professional skill than for his unassuming disposition and benevolence as a man, and for the part he played for upwards of a quarter of a century in connexion with gigantic public works.

Isambard Kingdom Brunel (1806–1859)

I have already mentioned Isambard Kingdom Brunel, the only son of Marc and Sophia Brunel, who was born at Portsea on 9 April 1806. Like his father, he possessed a vivid and vibrant personality. A born leader, he was ready to assert his leadership in any project he undertook and to sustain it by great self-confidence and an enormous capacity for hard work. In the end, his colossal talent and exuberant energy was cut short by his premature death at the age of 53. Isambard, who took after his father in many respects, was educated in Paris at the prestigious Lycée Henri-IV, at this period celebrated for its staff of mathematical teachers. His aim was to enter the Ecole Polytechnique but he was disqualified because of his foreign birth. Although his father wanted him to remain longer in France, nevertheless, Isambard returned home in 1822 at the age of 16 and began work in his father's busy office. There he became increasingly a partner of his father, who was involved in planning paddle tugs on the Rhine, a cannon-boring mill for the Netherlands government, a new bridge across the Serpentine, and other projects, including the Thames tunnel. At the age of 20, he was appointed resident engineer of the ill-fated project. Soon, as the result of an inundation, he was seriously injured and was not back at work again until the following year. By then work on the tunnel was at a standstill, owing to lack of funds, and was not resumed until seven years later and not completed until 1843.

Meanwhile, although he continued to live and work in London, Bristol began to occupy a good deal of Isambard's time and energy. He was originally drawn to the city by the competition for a road bridge across the Avon Gorge linking the city with Clifton. Later he was involved in improving the harbour, designing ships that were constructed there and above all created the Great Western railway between Bristol and London. The competition organized by the Clifton Suspension Bridge Company was adjudicated by Telford, who entered his own design after finding fault with the other entries, including Brunel's. However, Telford's design was rejected

by the trustees as too expensive and Brunel managed to modify his design sufficiently to overcome the objections to his proposal. He was thereupon appointed engineer-in-chief to the project but there were 20 years of delay, due to financial and legal difficulties. Brunel was employed in other projects in Bristol and elsewhere, and did not live long enough to see the Clifton Bridge finished.

For the transatlantic trade, Bristol was facing severe competition from Liverpool, which offered much better facilities. Sailing ships were beginning to be replaced by steamships, but only for comparatively short voyages. The Great Western and Great Britain, designed by Brunel, showed that steamships could successfully cross the Atlantic, and soon they displaced sailing ships on the crossing to New York, usually from Liverpool. The original purpose of the Great Western Railway was to provide a link between London and New York via Bristol, but the deficiencies of Bristol as a port meant that its main function was rather the transport of passengers. Brunel saw it as an express service, restricted to first and second class passengers. The construction of the railway was his greatest achievement as an engineer, occupying 20 of the best years of his life. The Great Western set a very high standard. Brunel designed the bridges and tunnels, including the Box tunnel outside Bath, and the splendid termini in London and Bristol. The opening of the line from London to Bristol in June 1841 was just the start of a network connecting other important places in the west of England and South Wales. When the line was extended into Cornwall, Brunel designed the famous Royal Albert Bridge across the Tamar, which is still in regular use. He also built the short-lived Hungerford suspension bridge in London, the wrought-iron chains for which were later transferred to the Clifton Bridge when work on that was resumed.

However, he made three serious mistakes. First, he chose the broad gauge, in preference to the narrow gauge, which was used on almost all the railways that had so far been constructed. There were good reasons for that decision but the inconvenience of the change of gauge was too great and eventually a parliamentary commission decided that the narrow gauge should be the industry standard, although the broad gauge was not finally abolished until 1892. The second was that the specifications he insisted on for the original locomotives made it impossible for the builders to meet his speed requirements. Later improved designs were to overcome this problem but it was strange that Brunel seemed unaware that it existed. The third mistake was to adopt the atmospheric or pneumatic system for part of the network, whereby stationary engines exhausted a tubular third rail into which the train inserted a piston. This simply could not be made to work satisfactorily, but if it had it would have provided many of the advantages that electric traction does today.

Brunel was employed as a consultant in railway construction in other countries, notably Australia, India, Ireland and Italy. His health had been declining in the mid-1850s. His kidneys were failing and he was advised to take a holiday, with his family, which they took in Egypt. However, the projects that dominated the closing years of his life were the design and construction of steamships for the Atlantic crossing. The first, the *Great Western*, was a fairly conventional timber-hulled paddle steamer, but her sister ship, the *Great Britain*, the largest vessel afloat, was the first to be iron-hulled and screw-driven. The third and last, the *Great Eastern*, a gargantuan iron ship, was far larger than any ship then in existence. It was intended to provide transport for up to 4,000 passengers between Britain and the Antipodes. This project amounted to an obsession; at the time of his death, the ship had been launched, with great difficulty, but he did not live to see its maiden voyage, which ended in disaster.

Rolt (1957), in his best-selling biography of Brunel, blames John Scott Russell for what happened to the *Great Eastern*. Emmerson (1977) has disputed this. Russell, a scientist, shipbuilder, naval architect, engineer and man of affairs, collaborated with Brunel in the design and construction of the great ship. It was a time of general financial pressure, felt by the shareholders and by both Brunel and Russell personally. When Russell's shipbuilding firm went into liquidation, Brunel took control of the launching of the great vessel. Russell lent a helping hand; Robert Stephenson also lent moral support. Russell's creditors helped him back into business, as a subcontractor rather than a partner in the *Great Eastern*. On its trial voyage, there was an explosion which killed five stokers; Russell was present, but not in charge; Brunel was at home, incapacitated by a stroke. The great ship never fulfilled its intended purpose, but was usefully employed in cable-laying and other tasks.

Brunel died on 15 September 1859, and was laid to rest in the family tomb in Kensall Green Cemetery. He had married Mary Elizabeth Horsely, a member of a sociable and artistic family, in 1836; she bore him two sons and a daughter. The elder son became an ecclesiastical lawyer, the younger an engineer. The sons deeply resented Russell's rôle in the *Great Eastern* affair, and their version of events was no doubt the source of Rolt's distorted account. Perhaps Emmerson's defence of Russell goes too far in the other direction; as Buchanan (1983 and 2002) has argued. When only 20, Brunel was elected a fellow of the Royal Society, like his father, and, like him, received an honorary degree from the University of Oxford. Although he purchased a country estate near Torquay in Devon he was too busy ever to go there.

John Augustus Roebling (1806–1869)

America needed civil engineers to build bridges across its great rivers. I profile two of them, one of German origin, the other native-born. John Roebling was born in Mühlhausen, a small trading city in Thuringia, on 12 June 1806, the son of Christopher Polycarpus Roebling, a retail tobacconist, and Friederike Mueller. He studied civil engineering at the Royal Polytechnic Institute in Berlin, where he was taught by the philosopher Hegel who thought highly of his ability. After he graduated in 1826, he was required to spend three years working for the bureaucratic and oppressive Prussian government. When this was over, Roebling migrated to the United States with an idealistic group from his hometown, some of whom set up a farming community they called Saxonburg, near Pittsburgh. There in 1836, he married Johanna Herting, daughter of another emigrant from Mühlhausen; they had nine children. After his wife's death in 1864, Roebling married Lucia W. Cooper; he had no further children.

Roebling soon became discontented with agriculture and wanted to return to engineering. He designed a steam plough, which never got manufactured, and in 1837 he went to work for a canal company in Western

Pennsylvania, which used winch-driven hemp ropes to haul canal boats up and down inclined planes when the canal they were travelling along came to a hill. However, the ropes sometimes broke under very heavy loads, and in 1841 he witnessed such an accident when two men were killed.

About this time, having read in a German technical journal about the possibility of producing wire rope, Roebling devised – and patented – a machine to make twisted wire cables strong enough to pull canal boats up a ramp without breaking. His wire ropes were so strong that Roebling used them between 1845 and 1850 to construct a number of suspension aqueducts for the Delaware and Hudson canal, eliminating the need to pull canal barges over hills. These wire-suspended aqueducts established Roebling's reputation as a builder. The demand for Roebling's wire ropes became so great that in 1848 he moved to Trenton, New Jersey, where his company, formed in 1842, became extremely prosperous, producing everything from chicken wire to massive cables capable of sustaining great weights.

Roebling's successful use of wire ropes for suspending canal aqueducts had taught him a great deal about the characteristics of suspension bridges, or chain bridges as they were originally called, and his mechanical ingenuity, structural imagination and entrepreneurial ability eventually made him the leading builder of such bridges in the United States. His first major success was his design for a double-decker bridge across the Niagara river at the Falls, with an upper deck for a railway and a lower deck for carriages. In 1851, Roebling began construction of the Niagara Falls Bridge, but he changed his original design three years later when another suspension bridge over the Ohio river at Wheeling, West Virginia – the largest span length in the world – collapsed. To avoid such a catastrophe befalling his Niagara Bridge, Roebling added trusses to stiffen and reinforce the heavy weight of the spans. This significant improvement made suspension bridge roadways rigid and impervious to varying loads as well as to destructive oscillations caused by high winds. Completed in 1855, Roebling's Niagara Bridge was the first successful railroad suspension bridge. It brought Roebling other contracts and encouraged him to make further innovations in construction.

In constructing another bridge over the Allegheny river in Pittsburgh, Roebling employed a new method of spinning cable on site, using travelling sheaves that moved on temporary cables over the support towers from one anchor to the other. He then used the same procedure on a bridge over the Ohio river at Cincinnati, with a 1057-foot span, a record for that time. In addition to stiffening trusses, this bridge had bound cable of parallel wires, wire rope suspenders and auxiliary stays radiating downwards and outward from the tops of the tower to the deck, thereby achieving additional aerodynamic stability. These features, which became characteristic of Roebling's bridges, also possessed the unique ability to stir the aesthetic feelings of those viewing the bridges. Although begun in 1856, Roebling's Cincinnati Bridge was not opened until 1867, as a result of delays caused by financial difficulties, the Civil War and natural disasters. A testimony to the bridge's strength was its use throughout the twentieth century.

The same holds true for the Brooklyn Bridge. The desirability of a bridge across the East River, linking Brooklyn with lower Manhattan, had long been recognized. In 1867, the New York State legislature chartered a private company to raise money for its construction. Roebling's successful record of building major bridges made him the logical choice of chief engineer to design and build it. The main span of the Brooklyn Bridge extends 1595.5 feet between the towers, and the two anchor spans each add another 930 feet. For extra strength, Roebling used steel, instead of iron, wherever metal was required. Since the towers had to stand on bedrock, far below the alluvial sands of the river bed, they were built on top of caissons made up of concrete and granite blocks weighing up to six tons each. When completed, the towers stood 273 feet above the water, the tallest structures in the country at that time.

However, Roebling did not live to see his design carried out. In June 1869, while he was inspecting a site for locating one of the towers, a ferry hit the dock where he was standing, smashing the toes of his right foot between two timbers. Although doctors amputated his crushed toes, some three weeks later he died of tetanus, before construction of the bridge was due to start. Roebling's 32-year-old son, Washington Augustus, took over as chief engineer. He had trained at the Rensselaer Polytechnic Institute, risen to the rank of colonel in the Union army, had collaborated with his father on several previous projects, and had just returned from studying new methods of bridge construction in Europe. He understood what needed to be done but before he could complete the project he became paralyzed as the result of compression sickness caused by inspection of the bridge's caissons. His capable and determined wife now took over and when it is said that the Roeblings built the Brooklyn Bridge, her contribution should not be overlooked. When she died in 1903, Washington Augustus married again, this time a southern belle named Cornelia Witsell Farrow.

At the time of its opening, the Brooklyn Bridge was the largest in the world. It had great economic, political, technological and aesthetic implications. By connecting Brooklyn with lower Manhattan, it created an economic surge in both communities, which helped to bring about the consolidation of five boroughs to create the metropolis of New York City. It had the strength to carry much more weight than originally anticipated and although some structural alterations were required in the 1940s it retained its basic structural and aesthetic elements as originally constructed.

Sir Joseph William Bazalgette (1819–1891)

Like Brunel and Vignoles, Bazalgette's forebears were French. The grandfather of the famous civil engineer was born in the small town of Ispagnac in the sparsely inhabited Massif Central in 1760. He moved to England after becoming wealthy in the West Indies. By his first marriage he had three children, of which the only son Joseph William was born in 1783. Joseph William made his career in the Royal Navy and in 1809 was wounded in action against the French. Five years later he retired from the navy with

the rank of commander and went to work for the Naval and Military Bible Society.

Joseph William married Theresa Pilton; their fourth child and their only son Joseph was born on 28 March 1819 at Enfield, Middlesex. At the age of 17, Joseph became an articled pupil of one of Telford's principal assistants, to be trained in civil engineering. Before long, he was employed as resident engineer on land drainage and reclamation work in Northern Ireland. By 1842, aged 23, Bazalgette had set up his own civil engineering practice, designing and laying out schemes for railways, ship canals and other engineering works in various parts of the United Kingdom and preparing the surveys and plans to meet parliamentary requirements. Before long, there was so much business that he was employing a large staff of engineering assistants. His office was near the Institution of Civil Engineers, of which he was already a full member and would later be elected president. The experience that Bazalgette had gained in preparing plans under the pressure of deadlines and in dealing with legal requirements during this testing period was to be invaluable in his later work.

Bazalgette was of somewhat delicate health – he suffered severely from asthma – and under the strain of the heavy work, his health completely gave way in 1847. He retired from business temporarily and went into the country, where a year of complete rest restored him to health.

The Metropolitan Board of Works had been established to modernize the infrastructure of a city that had hopelessly outgrown its medieval origins. In 1856, Bazalgette was appointed chief engineer of this powerful body, a post which he held for the 33 years of its existence. During that time he was responsible for executing well over 20 million pounds worth of work on behalf of the board. To start with, it was generally recognized that the sewage system of the metropolis was grossly inadequate. The waste of the city was emptied into the river Thames and its tributaries, from which drinking water was extracted. The heavily polluted river water was noisome and a major cause of disease, particularly cholera. By 1851, the broad outline of a scheme had been drawn up consisting of interceptor sewers to divert sewage from the Thames, with separate systems north and south of the river and with remote outfalls some 12 miles below London Bridge. The scheme had been proposed by Bazalgette's predecessor in office but it was left to Bazalgette and his assistants to carry it out. When it was finished in 1875, it comprised 130 miles of large-diameter sewers and four pumping stations.

As part of the main drainage scheme, Bazalgette fulfilled the long-cherished plan of embanking the Thames in central London. The Albert (1868), Victoria (1870) and Chelsea Embankments, of total length three and a half miles, reclaimed 52 acres of riverside land. Also, an Act of 1877 empowered the Metropolitan Board of Works to purchase the Thames bridges from the private companies that owned them and free them from tolls. Bazalgette decided that three of these bridges needed to be completely rebuilt, to meet the demands of modern traffic. The new bridges were the masonry arch structure at Putney (1886), the steel-link suspension bridge at Hammersmith (1887) and the iron arch structure at Battersea (1890). Three new river crossings were also constructed: the Tower Bridge, the Blackwall tunnel and the Woolwich free ferry; Bazalgette was involved in each of these projects although only the last was built to his design. At the same time, he also carried out the design and construction of new thoroughfares, of which the most notable were Southwark Street (1864), Queen Victoria Street (1871), Northumberland Avenue (1876), Shaftesbury Avenue (1886) and Charing Cross Road (1887). Several of these street improvements involved the clearance of slum properties.

Besides his work in London, Bazalgette was active as an engineer in other spheres and his reputation ensured that his services were frequently called upon by other communities in Britain and overseas. In 1865, he was invited to prepare a scheme to divert sewage from the river at Norwich; a similar scheme for Cambridge was not carried out. This was also the fate of his plans to clean up the river Danube at Budapest, but another scheme to provide sewers for Port Louis in Mauritius was carried out. However, it was in London that Bazalgette made his greatest contribution.

In 1870, Bazalgette was made a Companion of the Order of the Bath, for his services to London; four years later he received a much-deserved knighthood. He died on 15 March the following year at his home in Wimbledon, Surrey, survived by his Irish wife, Maria Keogh of Wexford, six sons and four daughters. A contemporary described him as very slight and spare, and considerably under the average height; but his face, with its prominent aquiline nose, keen grey eyes, grey whiskers and black eyebrows, gave the impression of a man of exceptional power.

His obituary in the *Proceedings of the Institution of Civil Engineers* (ICE, 1891) stresses the personal qualities that enabled him to carry out his great work in the face of difficulties that others would have found insurmountable. The technical and political problems that accompanied his work demanded great patience and the use of new materials. The changes he made to the face of London in the form of new bridges, parks and streets, and the railways and sewers running below them, have remained to the present day. He was not an entrepreneur in the heroic mould like Brunel but he had to exercise heroic patience in dealing with politicians, vestrymen and Board members.

5 From Eads to Bell

James Buchanan Eads (1820–1887)

The name of Eads has already been mentioned in the profile of Roebling. James Buchanan Eads, the future bridge-builder, was born in Lawrenceburg, Indiana, on 23 May 1820, the son of Thomas Clark Eads, a businessman, and Ann Buchanan. He was named after his mother's cousin, James Buchanan, a Pennsylvania Congressman who later became the undistinguished 15th President of the United States. After leaving school at the age of 13, he tried his hand at a variety of jobs, until in 1838 he became purser on a Mississippi steamboat, the *Knickerbocker*, which before long hit a snag in the river and sank, like many other boats. Eads decided to become a salvage engineer, using his own design of diving bell and by the time he was 25 he had made enough money to invest some of it in establishing the first glassworks west of the Ohio river. When this proved unsuccessful he returned to salvage work on the Mississippi until 1857.

At the outbreak of the Civil War, in 1861, Eads realized that control of the Mississippi would be an important strategic objective and advised President Lincoln to acquire a fleet of steam-propelled iron-clad gunboats to patrol the western rivers. Although he had no experience of shipbuilding he successfully contracted to build seven such boats without their guns in 65 days. After these gunboats provided the first Union victories of the war, he was awarded a second contract to build or convert another 18.

In 1866, when the war was over, he successfully bid for the contract to construct a bridge across the Mississippi at St Louis. He hired German-trained engineering assistants to calculate the structural plans and materials, so that he could make realistic cost estimates. He knew so much about the currents and the peculiarities of the river bed that he was successful in bridging the great river with three spans of over 500 feet. The superstructure was made of cast iron, the tubular arches of steel. To provide it with adequate foundations, he used pneumatic caissons, a method developed in Europe but not previously used in the United States. Because the physiological dangers of working under high pressure were not well understood, his workforce suffered the sickness known as the bends, and some of them

died as a result. Eventually, it was realized that slow decompression was the answer to this problem but Roebling did not know this.

After seven years of work, the Eads Bridge, as it is known, was finished triumphantly in 1874. The following year, Eads was commissioned to open one of the channels in the estuary of the Mississippi, which he accomplished successfully. His last major project was to try and construct a ship railway across the Isthmus of Tehuantepec, instead of a canal in Panama. This did not get very far but the tropical conditions undermined his health, and he died at Nassau in the Bahamas on 8 March 1887 at the age of 67. Eads was married twice. He had three children by his first wife, one of whom died in infancy. After she died of cholera, he married a widow with three children to add to the two surviving children of his first marriage, but there were no more children. The Eads Bridge at St Louis still stands, carrying heavy railroad traffic on one deck, highway and streetcar traffic on the other. Eads was the first American to be awarded the Albert Medal of the Royal Society of Arts for his extraordinary achievements.

William Thomson (Lord Kelvin of Largs) (1824–1907)

We now come to a man who was mainly a physicist for the first half of his long life, mainly an engineer for the second. Of course, the two activities were closely linked. William Thomson, the future Lord Kelvin, was born on 26 June 1824 in a comfortable house on the outskirts of the Irish city of Belfast. He was the fourth of seven children, four sons and three daughters, their mother Margaret, née Gardner, came of a Scottish mercantile family. Their father, James Thomson, an Ulster Scot, was professor of mathematics at the non-sectarian Royal Belfast Academical Institution. Margaret Thomson died in 1830 shortly after giving birth to their youngest son Robert, so the upbringing of the children devolved directly on the father. In 1832, the family, reduced to six children by the death of one in infancy, moved to Glasgow, where James Thomson became professor of mathematics at Glasgow College, the educational heart of the ancient University of Glasgow, where he had earlier been a student. He was already known as the

author of several mathematical textbooks, the royalties from which helped supplement his meagre salary.

Initially, William and his elder brother James were taught at home. In 1834, both boys began their formal education by matriculating at Glasgow, where natural philosophy was an essential part of an all-round course that began with philosophy and ended with theology. William Meikelham, then holder of the chair of natural philosophy, had a great respect for the French approach to physical science. Encouraged by him, William Thomson read Laplace's *Mécanique Céleste* and Fourier's *Théorie Analytique de la Chaleur* in French during a family continental tour in 1839. His brother James, who had a passionate interest in all things mechanical, had decided to become an engineer. The third son, John, was embarking on a career in medicine but succumbed to the epidemic of typhus in 1847, which followed on the famine in Ireland; the youngest brother, Robert, went into the insurance business and emigrated to Australia. Since William was most attracted to mathematics, his father sent him to Cambridge.

William Thomson was admitted to Peterhouse where he was coached by the famous William Hopkins. He participated fully in student life, winning trophies for sculling, and became one of the most popular undergraduates of his year. When the time came for him to sit the Tripos examination in January 1845, Thomson was disappointed to come out only second on the list; he expected to be senior wrangler; no one was in any doubt that he was the best mathematician of his year. However, in the highly competitive examination for the Smith's prize, which followed the Tripos, he came out first, having found the questions more to his liking. He was elected a foundation fellow of Peterhouse and college lecturer in mathematics later in the year, although still barely 21. He had done very well, but his future career was to be in Glasgow, not Cambridge.

Soon after graduation, encouraged by his father, he travelled again to France, and this time stayed much longer. Thomson's studies in Paris were crucial to the subsequent development of the British school of physics, but it was not so much the lectures as the practice in new experimental work that he found so important later. He was particularly indebted to his teacher, Victor Regnault, for a faultless technique, a love of precision in all things and the highest virtue in an experimenter – patience. During this period, he developed the technique of electrical imaging, read about Carnot's theory of the motive power of heat and formulated a methodology of scientific explanation that would strongly influence James Clerk Maxwell.

After his 1845 visit to Paris, Thomson returned to Scotland. The next year, following the death of Meikleham, he was elected to the professorship of natural philosophy in the University of Glasgow, the post he held for the remaining 53 years of his life. The election was hotly contested but in the end he was elected unanimously. He was also elected a Fellow of the Royal Society of Edinburgh. In 1849, Thomson's first great memoir, *A Mathematical Theory of Magnetism*, was published, which led to his election to the Royal Society of London; later numerous other learned societies elected him to membership. In 1851, a second great memoir appeared, *On the Dynamical Theory of Heat*. In this work, he postulated the existence of a state of complete rest, which he called absolute zero, the base of the temperature scale to which the name Kelvin is nowadays attached. In these years he used to spend some time each summer in Cambridge, to keep in touch with the scientists there and in London, and once or twice he also revisited Paris.

At that time nowhere in Britain was there a research laboratory in a university or anywhere else where students could work. Thomson, having enjoyed such facilities in Paris, was keen to establish similar opportunities for his students, so he extracted a small sum from the university for that purpose. It made possible the first teaching laboratory in Britain, albeit of a modest sort. He was also greatly interested in developing measuring instruments of high accuracy, and the facilities of the new laboratory made that possible too. The new professor soon became popular through his radically new professionalism, marked by a clear research orientation.

For an impression of Thomson in the lecture room we have these lines by one of his students, written years afterwards.

> Lord Kelvin, as Thomson later became, possessed the gift of lucid exposition in ordinary language remarkably free from technicalities. Occasionally he got out of range of the majority of his class, but here was no obscurity in his statement, it was simply beyond their grasp ... he had no set syllabus and used no notes in lecturing. He had his subject clearly before him and dealt with it in logical order. He was not dictating a manual of natural philosophy to his students ... he considered it unnecessary for him to teach what could be got in an ordinary textbook and that his province was to supplement this. On one occasion he described his method as follows: the object of a university is teaching not testing. The object of the examination is to

promote the teaching. The examination should in the first place be daily. No professor should meet his class without talking with them. He should talk with them and they with him.

After his beloved father had died in an outbreak of cholera in 1849, Thomson began actively seeking a wife. Disappointed in his overtures to one young lady, he married, on the rebound, the highly cultured and intellectual Margaret Crum in 1852. She was three years younger than he was and came from one of the newer commercial families of Glasgow, with whom he had been friendly for many years. Unfortunately, her health declined rapidly following an exceptionally strenuous honeymoon tour around the Mediterranean, and attempts at treatment were largely unsuccessful. It is not known what was wrong with her but after prolonged suffering and many relapses she died in 1870 after 17 years of marriage.

The rôle that the engineer James Thomson played in the life of his younger brother must not be underestimated. After practising as a civil engineer in Belfast for some years, James had become professor of civil engineering at Queen's College, until, in 1837, he moved to be professor of engineering at Glasgow. The vortex turbine was his speciality. The Cambridge mathematical physicist Sir Joseph Larmor referred to James as the philosopher who plagued his pragmatic brother to obtain a comprehensive understanding of the problems he dealt with. As the German scientist Hermann von Helmholtz explained:

> James was a level-headed fellow, full of good ideas, but cares of nothing except engineering, and talks about it endlessly all day and all night, so that nothing can be got in when he is present. It is really comic to see how the two brothers talk to one another, and neither listens, and each holds forth about quite different matters. But the engineer is the more stubborn and generally gets through to his audience.

At dinner parties where both brothers were present it was considered advisable to seat them as far apart as possible.

For a number of years, Thomson collaborated on research with Carnot's self-educated English follower James Prescott Joule. In a posthumous tribute to Joule, Thomson said that:

> The genius to plan, the courage to undertake, the marvellous ability to execute and the keen perseverance to carry through to the end the great series of experimental investigations by which Joule discovered

> and proved the conservation of energy in electric, electromagnetic and electrochemical actions, and in the friction and impact of solids and measured accurately, by means of the friction of fluids, the mechanical equipment of heat, cannot be generally and thoroughly understood at present. Indeed it is all the scientific world can do just now in this subject to learn gradually the new knowledge thus gained.

Telegraphy was by this time a well developed and extremely profitable business, and by 1850 there was already a successful submarine cable between England and France. However, it was the Atlantic cable project of 1866 that most caught the popular imagination. Thomson was brought in early as a member of the board of directors of the project, and he played a central rôle in its execution. The directors had entrusted the technical details to an industrial electrician named Whitehouse, and many of the difficulties that plagued it from the outset resulted from Whitehouse's insistence on employing his own system of electric signalling. Thomson had developed a very sensitive apparatus, the mirror galvanometer, to detect the minuscule currents transmitted through the miles of cable, but Whitehouse refused to use it. Thomson predicted that the length of cable would, by a process of statical charging of the insulation, substantially reduce the rate at which signals could be sent unless low voltages were used, so low that only his galvanometer could detect the currents.

The first attempt to lay such a cable in 1857 ended when it snapped and was lost. The second attempt a year later was successful, but the high voltages required by the Whitehouse method reduced the ability of the cable to transmit signals rapidly, just as Thomson had predicted. Whitehouse privately admitted the inadequacy of his own instruments and surreptitiously substituted Thomson's galvanometer while claiming success for his own methods. This deception was soon discovered and the ensuing controversy among Whitehouse, the board of directors and Thomson combined theoretical science, professional vanity and financial ignominy. A third cable was laid in 1865 and, with the use of Thomson's instruments, it proved capable of rapid transmission. Thomson's rôle as the man who saved a substantial investment made him a hero in the British financial community and to the Victorian public in general; indeed he was knighted for it by Queen Victoria in 1866.

Sir William's success with the Atlantic cable and the close relation thereby established between testing in the laboratory and application to the electrical industry opened the way for the ambitious marketing of scientific knowledge: through a carefully developed and cleverly exploited system of patents and partnerships his financial returns on his scientific capital were such that he soon became a wealthy man. This kind of success was deplored by other scientists, especially those in France and Germany, who maintained an almost religious belief in the importance of keeping science pure, uncontaminated by industrial applications.

Some of his most useful and profitable inventions were related to navigation. One of the most successful was the patent magnetic compass of which no fewer than 10,000 were sold; this eventually superseded the older type, which was unreliable in iron-clad ships. Sir William purchased a schooner of 126 tons partly for his own pleasure but also so that he could try out such inventions at sea. It also provided a way in which he could escape from the pressures of his many different responsibilities. One cruise took him to Madeira, where he was entertained by the Blandy family of wine shippers. It was then that he met Frances Anna Blandy, who was about 14 years younger than he was, and they were married in 1874. Neither of his marriages resulted in children.

Throughout the latter part of the nineteenth century the progressive University of Glasgow was becoming increasingly cramped on its ancient site among the slums of the High Street. Eventually, enough finance was raised by public subscription and by the philanthropy of local industrialists to build a grand new cathedral of learning on a new site, which opened in 1870. It included a new physical laboratory, vastly superior to the old one. In 1891, the students elected Sir William Lord Rector; in 1904 he was appointed Chancellor of the University.

In 1876, the science journal *Nature* printed an appreciation of Sir William's achievements by Helmholtz, which concludes by saying that:

> British Science may be congratulated on the fact that in Sir William Thomson the most brilliant genius of the investigator is associated with the most lovable qualities of the man. His single-minded enthusiasm for the promotion of knowledge, his wealth of kindliness for younger men and fellow-workers, and his splendid modesty, are among the qualities for which those who know him best admire him most.

In 1892, Sir William was created Baron Kelvin of Largs in the county of Ayr, the first scientific peer of the realm (the Kelvin is the river that runs through the gardens of the University of Glasgow, and Largs was the nearest town to Netherhall, where the Kelvins lived). In 1896, the jubilee year of his professorship, he was also decorated with the Grand Cross of the Royal Victorian Order and in 1902 became one of the first members of the Order of Merit and a member of the Privy Council. In 1899, the year in which he retired from his chair, he had become the first foreigner to be decorated with the Grand Cross of the Legion of Honour. His eldest brother James died in 1892, at the age of 70, and Elizabeth, the eldest sibling and last surviving sister, also died in the same year.

Kelvin enjoyed good health throughout his life, apart from a fractured thigh-bone caused by a fall when he was engaged in the Scottish sport of curling on ice. In September 1907, he suffered a major stroke; at the end of November he became gravely ill and he died on 17 December at the age of 83. The funeral took place in Westminster Abbey, two days before Christmas.

Gustave Eiffel (1832–1923)

Born in Dijon, the provincial capital of Burgundy, on 15 December 1832, Gustave Eiffel was the first child and only son of his parents; later they

had two daughters, Marie and Laure. On his father's side, his forebears were German master-weavers who had moved to France early in the eighteenth century and established a successful tapestry studio in Paris. His father, however, had a military career, which brought him to Dijon, where he left the army and married Catherine-Melanie Moneuse, daughter of a timber merchant. In contrast to her self-effacing husband, she was a spirited and strong-willed woman who established a business supplying blast furnaces with coal transported by barges on the new Canal de Bourgogne. Throughout their marriage, she was the dominant parent, largely responsible for the family's prosperity.

Thus, Gustave Eiffel was a native of Burgundy, like Marc Brunel. He attended the Lycée Royal in Dijon but his education did not really begin until he went to Paris to study at the Collège Sainte-Barbe. Although he might have qualified for the Ecole Polytechnique he decided instead to choose the more vocational Ecole Centrale des Arts et Manufactures, where chemistry was his speciality. Contemporaries described him as a rather prim, timid and conventional student, socially rather dull. After graduating from the Ecole Centrale, he sought to obtain some practical experience of industry and was eventually given a junior position in La Compagnie des Chemins de Fer de l'Ouest, the owner of some of the principal French railway lines.

The company's chief engineer gave him some excellent training, especially in the structural use of iron and steel, and before long he was able to design a small railway bridge using these materials rather than the traditional timber and masonry. When control of the company changed hands, Eiffel was promoted to a much more senior position and given his first major project, the design and construction of a 1,640-foot iron railway bridge across the Garonne at Bordeaux. He proved himself an astute manager and an inspired organizer.

Now that Eiffel's prospects seemed good, he proposed marriage to a childhood sweetheart from Dijon named Marie Gaudelet. They were duly married in 1862 and settled down in a large house in the north of Paris. She gave him two sons and three daughters before she died in 1877.

The company Eiffel worked for was forced into liquidation, but before that happened he set up as a consulting engineer on his own account. He was fortunate to be given an order by the Egyptian government to supervise the building and delivery of 33 locomotives. This meant that he needed to visit Egypt, where he met Ferdinand de Lesseps (1805–94), then engaged in the construction of the Suez Canal.

Eiffel planned to benefit from the rapid expansion of the French railway network, which was taking place. So that he would be better able to compete for construction contracts, he purchased substantial iron-working premises in Paris. Soon his business was thriving. His reputation soared with the 1,864-foot railway viaduct he constructed at Garabit, south of Clermont-Ferrand. This rose 400 feet above the valley floor; at the time that it was built, it was the highest arched bridge in the world, and the most graceful of all Eiffel's bridges. It was constructed using techniques that he was to employ later to build his celebrated tower. Another remarkable achievement was the 1,882-foot Cubzac road bridge in the Dordogne. For his work in France, Eiffel was awarded the Grand Cross of the Legion of Honour, but he also completed major projects in other countries; a new railway station in Budapest and a bridge across the river Douro at Porto in Portugal.

While Eiffel was working on the Garabit viaduct, he was consulted about another project which was very much in the public eye. In 1870, liberal republicans at a dinner in Paris had discussed the idea of presenting a statue to France's sister republic on the other side of the Atlantic. The project gathered momentum and by 1884 the young Alsatian sculptor Frédéric Auguste Bartholdi had designed a colossal 155-foot statue symbolizing *Liberty Enlightening the World* to be placed on a 150-foot plinth inside the courtyard of Fort Wood on Bedloes Island at the entrance to New York harbour. The cost of the statue itself was raised by public subscription and a lottery in France; while the cost of the plinth was raised, not without difficulty, in America. Eiffel was consulted about engineering problems, such as wind resistance. The statue was shipped across the Atlantic in pieces and after it was erected the world-famous monument was formally dedicated before an enthusiastic crowd on 26 October 1889.

Although Eiffel did not play a leading rôle in this project, it brought him valuable publicity. As a result, his name was put forward to design a centrepiece for the international exhibition to be held in Paris in 1889. The French government invited proposals for a 300-metre iron tower with a base 125 metres square to be erected on the Champ de Mars. Eiffel submitted his proposal although he was not personally involved at this stage; the design owed more to his assistants. There was strong opposition from the art establishment, who wanted the tower built of stone, or in a different location, or not built at all. In the end, however, Eiffel's firm won the competition and he at once took charge of the project. To secure the necessary finance, he assumed a greater part of the risk personally, in return for a share of any profits. It was completed without any major problems; the only casualty among the workforce was one man who, after work, was showing off by climbing the exterior girders of the tower rather than using the staircase. From the day it was opened to the public in the spring of 1889, it was and continued to be a huge success, drawing almost 12,000 visitors a day. This made Eiffel a very wealthy man; there was no further need for him to work for a living.

Another prestigious project being started at this euphoric time was the attempt to construct a canal across the Isthmus of Panama. Shares in the enterprise were offered for public subscription at the end of 1880. Lesseps, who had brought the construction of the Suez Canal to a triumphant conclusion, was put in charge of the ill-fated project. Geographical conditions and disease held up construction; further finance was desperately needed. Eiffel was drawn in as a consultant when the design was changed from a level waterway to one involving a system of ten huge locks. There was a high degree of mismanagement, but it was assumed that the French government would never allow the project to fail. Eventually, the company was declared bankrupt and the directors indicted on charges of fraud and breach of trust. While Eiffel was not a director, he was included in the indictment; and although acquitted of fraud he was found guilty of breach of trust and sentenced to two years imprisonment, suspended pending an appeal. The verdict was confirmed by the court of appeal but quashed by the supreme court in 1894. Work on the canal was abandoned and it was only much later that the American government revived and completed the construction.

Eiffel's reputation had been tarnished and he was never again to complete a major project. He proposed an electric railway, partly underground and partly overground, around the centre of Paris. Competing proposals and indecision meant that nothing was done for the next ten years. He also studied the old problem of constructing a crossing of the English Channel. Some favoured a bridge, which would cause navigational problems; others, of whom Eiffel was one, favoured a tunnel. Again nothing was done. Finally he was persuaded to design and build an astronomical observatory on the summit of Mont Blanc. Work on this was begun but stopped after an accident. Later a wooden building was constructed but abandoned after being crushed by ice movement. He died peacefully at his Paris mansion on the Rue Rabelais on 27 December 1923, at the age of 91.

George Westinghouse (1846–1914)

Although some American engineers were immigrants, others were native-born. George Westinghouse was one of these. He was born on 6 October 1846 in the farming village of Central Bridge, in the north-east corner of New York State. His father, of the same name, came of Saxon stock; his mother Emeline, née Vedder, was of Dutch-English stock. They had six children, of which two died in infancy. In 1856, his father, who was mechanically minded, gave up farming and moved the family to Schenectady, in the State of New York, where he formed a company manufacturing small steam engines and farm implements. George spent his spare time in the workshop, where he invented a rotary steam engine before he was 15 years old. He had a brother, Henry Herman, who later invented a high speed engine suitable for powering dynamos.

At the age of 16, George, with two of his brothers, ran away from home and enlisted in the New York National Guard, a kind of militia, hoping to fight in the Civil War, but was recalled by his parents because he was underage. He persuaded them to let him join the Cavalry but after a year transferred to the Union Navy, where he served as an engineer. When the war was over, he spent a short while at Union College in Schenectady before joining his father's firm. At the age of 20, he took out the first of many

patents, for the rotary steam engine he had designed earlier and for a simple device to facilitate the replacement of derailed railroad cars on the tracks. The latter was commercially successful although the former was not. He then moved to Pittsburgh, where he patented his first major invention, a compressed air braking system, which allowed all the individual brakes on trains and other vehicles to be applied at the same time. Previously, it was necessary for a brakeman on each pair of carriages to apply the brakes by hand when signalled to do so by the engine driver; this procedure was almost useless in emergencies and frequently led to accidents. Others had similar ideas but the Westinghouse design was far more practical; it enabled trains to travel safely much faster than before. After it was adopted by many American railroads, Westinghouse decided to investigate the situation in Britain. After initial scepticism, his braking system was demonstrated successfully but trains were already equipped with vacuum brakes and this stood in the way of any change. On the European continent, however, he met with greater success, and by 1881, Westinghouse air brakes were in widespread use throughout Europe as well as the United States. Westinghouse went on making mechanical improvements, which greatly facilitated the expansion and development of the American railroad system.

In 1865, Westinghouse had met by chance a young woman named Marguerite Erskine Walker on a train back from New York and decided there and then that he was going to marry her. They became engaged in due course, and were married in Brooklyn the following year; they had one son. In 1871, he bought a substantial house with some land at Homewood on the eastern outskirts of Pittsburgh, to which his wife gave the name Solitude. Large deposits of natural gas had recently been discovered nearby and Westinghouse decided that he would drill a well there for himself. When the operation was successful, he formed a company to supply natural gas to every house and small industrial plant in Pittsburgh. In the later part of his life, his family lived in Washington on Dupont Circle during the season, and they also had a country house, Erskine Park, at Lenox, Massachusetts.

Westinghouse began making experiments with electric railways for local traffic. This led him to develop the transmission of high-tension alternating electric current. Hitherto, everything electrical had used low-voltage direct current, the sort that was produced by chemical storage batteries. This had various drawbacks; power fell off rapidly with distance and it was not easy to step up or step down the voltage. High-tension alternating current did not have these disadvantages but there were concerns about its safety, following accidents where people were electrocuted.

By skilful purchase of patents and engineering skills, Westinghouse formed the Westinghouse Electric Company, based at Pittsburgh. Using technology invented by Tesla, whose profile follows later, he secured for his company the contract to illuminate the Chicago World's Fair and then, most importantly, to install the machinery in the huge hydro-electric power station at Niagara Falls.

By 1889, the Westinghouse Electric Company operated around the globe with a very large number of employees. Westinghouse developed a reputation for paying fair wages and providing generous benefits, so that he was able to avoid the strikes that plagued his competitors. Unfortunately, the company became heavily indebted and in 1907 he lost control in a financial crisis. The experience embittered him, although he continued to play a managerial rôle for the next four years. He also retained control of his Air Brake Company and successfully reorganized the Equitable Life Assurance Society when it was in trouble.

In 1911, Westinghouse reduced his business commitments; two years later his health broke down and he died in New York City on 12 March 1914. Honours he received included the Edison gold medal of the American Institute of Electrical Engineers; he was also awarded the Legion of Honour from France, the Order of the Crown from Italy and the Order of Leopold from Belgium. Unlike Edison, he never became a popular hero, the subject of much myth-making. There are dozens of books about Edison's life and work, but only two biographies of Westinghouse, neither of them at all recent. This is partly because he produced few truly original inventions in the course of his career; his strategy was more to take the ideas of others and develop them into commercial propositions.

Thomas Alva Edison (1847–1931)

Another native-born American engineer was Thomas Edison, who was born in Milan, Ohio, on 11 February 1847. He was descended from a Dutch miller, who came to America about 1735, but his mother, Nancy, née Elliot, was of Scottish-Irish descent. His father's family were landowners in New Jersey before the War of Independence, after which their land was confiscated and they moved with other loyalists to Canada. In 1837, however, his father had moved back to the United States, and in due course sent for his wife and children to join him in Milan. Thomas was the last of their seven children, three of whom died in childhood. He received a basic education from his mother, a former schoolteacher. Milan was a grain port, connected to Lake Erie by a canal. When Thomas was seven, the family moved to nearby Port

Huron, a lumber town, which was twice the size of Milan. Samuel Edison ran a grocery store, which was not a success, and he also farmed a truck garden. At the age of ten, his son was dividing his time between private experiments in chemistry and the sale of vegetables and newspapers; two years later he was employing other boys to do the work.

In his mid-teens, Edison left home and began working as a telegraph operator. Signals were sent in the dots and dashes of the Morse code. Although Morse invented this, and introduced practical telegraphy to the United States, telegraphic communication using electricity was not a new idea. In Europe, it was already being exploited commercially by the British scientist Sir Charles Wheatstone from about 1837; by 1850 there was a successful submarine cable between England and France. The telegraph and railroad companies for which Edison worked were the first great American corporations. The deafness from which he was beginning to suffer was perhaps an asset for this kind of work, since it helped him to concentrate. He moved around in the mid-west, earning just enough to survive, and began to think of ways in which telegraphy could be made more efficient.

In 1869, Edison moved to New York City, the headquarters of Western Union and other telegraph companies. It was also a centre for venture capital and finance generally. Although there was no immediate prospect of a job for him at Western Union, he was appointed superintendent of Law's Gold and Stock Reporting Company, which was about to be taken over by the larger firm. However, he was dismissed when this happened since Western Union already had a superintendent and did not need another. He went into business himself with a partner named Franklin Pope. Their company offered a variety of services, including instrument and wire testing, telegraph line construction, repair and maintenance, and patent application and drawings preparation. The partners devised a simple new printing telegraph marketed to merchants and importers who needed gold and other price quotations. Before long, Edison was hired by Western Union as consulting electrician and technician to develop simpler machines for specific telegraphic purposes. He insisted that the contract that he signed included the provision of workshop facilities.

Edison had exceptional ability to win the confidence of potential backers. Although he worked under contract to corporate sponsors he also worked independently He was unconcerned about conflict of interest and was inclined to reserve good ideas for himself rather than any of his sponsors. This gained him a reputation for unreliability. Orton, the president of Western Union, said that he came to believe that Edison had a vacuum where his conscience ought to be. He was not the only one; Jay Gould, the owner of the Atlantic and Pacific Telegraph Company, was another. Gould was fighting for control of Western Union and began by persuading Edison to come over to his side. When Bell successfully demonstrated his invention of the telephone, Edison immediately set to work to improve it. Once he had done so, Western Union started to fight the Bell interests, in what proved to be a lengthy contest. Eventually, it was agreed that Western Union would leave telephony to Bell while Bell would leave telegraphy to Western Union. The rivals fought it out in England too; the government intervened and forced the two sides to merge.

In 1871, Edison's mother died, and not long afterwards his father married again. On Christmas Day of the same year, the impulsive Edison married a 16-year-old working-class Newark girl named Mary Stillwell after a courtship lasting no more than eight days. She showed no interest in the work that preoccupied her husband. When he went to England on business, he made contact with the sophisticated British electrical community, led by William Thomson, later Lord Kelvin, and learned of new research on electrical transmission. He realized that there was much he did not know and transformed his Menlo Park workshop into more of a scientific laboratory. At the same time he began to assert his independence from his commercial sponsors.

Edison's Menlo Park laboratory has long been recognized by historians as providing a new model that helped transform American invention, and Edison himself has been described as a transitional figure standing between the lone inventor of the nineteenth century and the industrial researcher of the twentieth. Working in a tradition of cooperative shop invention, neither Edison nor most of his contemporaries were lone inventors, although they were generally independent inventors. The creation of the laboratory itself, however, was made possible by the growing interest of large-scale technology-based companies, such as Western Union, in acquiring greater control over the inventive process by supporting the work of these inventors. Although constrained by nineteenth century beliefs in the unpredictability of inventive genius, Western Union president William Orton clearly perceived the Menlo Park laboratory as an extension of the inventor himself. But he also recognized that the laboratory enabled Edison to make invention a more regular and predictable process and thus was willing to provide direct support for it. The support Edison received for his laboratory from Western Union and later from the Edison Electric Light Company helped to demonstrate the value of invention to industry and showed that invention itself could become an industrial process.

Most of Edison's inventions were simply improvements on existing devices, not involving any new principle. For example, there was his stencilling system for making facsimile copies, which survived for many years under the name of mimeography. He did not invent the typewriter, but tried in vain to make the invention a commercial proposition. He observed the effects of electromagnetic waves, well before Heinrich Hertz, but did not recognize their significance.

However, some of Edison's best inventions were of historic importance. The phonograph, as he called his talking machine, amazed the public and scientific community because of its utter simplicity. Edison had been seeking some way to store telegraph messages. In 1877, he had designed a device, which worked the first time it was tried. He had already recorded a message which, to the amazement of those present, inquired as to their health, asked how they liked the phonograph, informed them that it was very well, and bade them a cordial goodnight. Within a few months, Edison had become world famous. This is my baby, he told a reporter, and I expect it to grow up to be a big feller and support me in my old age.

Edison's efforts to realize the commercial potential of the phonograph were supported by three different groups of investors. His main support came from the Bell telephone interests, led by Gardiner Hubbard. As he had done with all his inventions, Edison reserved the foreign rights for himself. He appointed agents to ensure that foreign patents were obtained and that the invention was efficiently exploited by local companies throughout the world. The work of other researchers soon came to light and Edison was involved in controversy over priority, particularly over the design of the microphone. He preferred to avoid litigation, as far as possible, because of the demands it would make on his time.

Edison next turned his attention to electric lighting. At this time, lighting in towns with a gas supply was provided by gas lamps; elsewhere by kerosene lamps or candles. While Edison the individual is celebrated as the inventor of incandescent electric light, it was the less visible corporate organization of the laboratory and business enterprise that enabled him to succeed. His proven record as a successful inventor enabled him to raise the funds he needed for this work. Other experimenters had preceded him and there was plenty of competition but he was the first to produce a practical design that was commercially viable. In 1881, he shifted his workbase, and his home, from Menlo Park to Manhattan and the next year the system he installed in the offices of the financial house J.P. Morgan was in operation.

One of these conflicts was with the British inventor Joseph Swan, who had first experimented with incandescent lamps in 1848, when Edison was just a baby. To avoid an expensive patent dispute, he merged his company with Edison's British company in 1883. In 1884, the Edison electric light company began to alter its corporate strategy, confident that the Edison master patent of 1880 placed the company in a strong position to threaten infringement proceedings against its competitors. Unlike Westinghouse, Edison used direct current exclusively but, after a long struggle, alternating current became the industrial standard for most purposes, as it is today.

Edison's wife was suffering from ill health, following her third pregnancy. To improve her health Edison took her to Fort Myers on the Gulf coast of Florida to escape the worst of the winter, when she had him all to herself instead of just seeing him on Sundays. However, Mary died in 1882 leaving him a widower with three young children to bring up. A daughter, Marion, was the first, a son, Thomas Alva Jr, the second, and William Leslie the third. Marion showed real ability and assisted her father in various ways but Thomas, an alcoholic, endeavoured to trade on his relationship to his father.

Four years later he was married again, to Mina the daughter of Lewis Miller, partner in a firm that manufactured agricultural machinery from Akron, Ohio. Socially, the upper-middle-class Mina was a long way above the working-class Mary. They settled down in Llewellyn Park, a few miles from Newark, where Edison bought a large house named Glenmont. He developed a new laboratory nearby, to replace the one at Menlo Park, which had been seized by creditors. They started a new family; eventually this consisted of a daughter, Madeleine, and two sons, Theodore and Charles.

At the laboratory, Edison began by improving the gramophone to the point where it was commercially marketable. He then became interested in motion pictures. The relevant technology was not invented by him but his company supplied the projectors and films. From 1879, Edison also became seriously interested in mining, using electromagnets to concentrate the low-grade iron ores to be found in New Jersey, but in the face of competition from the rich and cheap iron ores of Lake Superior his mining enterprises were a failure and he lost both money and reputation as a result. By 1927, his business career was essentially over. He had become increasingly deaf. For some years he had suffered from diabetes and stomach disorders, which he treated by adopting a diet consisting of nothing but milk. In 1928, he was awarded the Congressional Medal of Honour. On 18 October 1931 he died, with his family around him. On the evening of his funeral, the nation, at the request of President Hoover, dimmed all electric lights for one minute.

The career of Thomas Edison was not that of a great man of science, or even that of an inventive genius whose path is illuminated by flashes of inspiration. In 1911, his nomination as a member of the National Academy of Sciences received only three votes. Joseph Henry described him as 'the most ingenious inventor in the United States or any other country'. His only major scientific discovery was the fact that the vacuum lamp could act as a rectifier, passing only negative electric currents. He failed to realize its potential, but later vacuum tubes were an essential part of radio sets. At the time of his death, Edison held 1,093 patents, the largest number ever awarded to a single individual. He was said to have invented the business of invention.

Conot (1979), in his well-researched biography of Edison, summed him up as a lusty, crusty, hard-driving, opportunistic and occasionally ruthless Midwesterner, whose Bunyanesque ambition for wealth was repeatedly subverted by his passion for invention. He was complex and contradictory, an ingenious electrician, chemist and promoter, but a bumbling engineer and businessman He was little motivated by the love of mankind but rather stimulated by a drive for self-expression, a desire for wealth and fame and a will to dominate. In America he is regarded as a prime example of the self-made man. 'I never did a day's work in my life,' he said: 'It was all fun.'

Alexander Graham Bell (1847–1922)

Alexander Graham Bell, the inventor of the telephone, was a Scot who moved to America as a young man. His paternal grandfather, of the same name, was a man of many parts, who began work as a shoemaker, then became an actor, ran a tavern, taught at St Andrew's Grammar school and, most significantly, practised as a tutor in elocution at nearby Dundee. The correction of stammering was a speciality of his. When his wife, Elizabeth, was unfaithful with the Rector of the Dundee Academy, Bell divorced her. In the settlement, she kept their 11-year-old daughter while he took their

14-year-old son Alexander Melville (1819–1905) with him to London. It was this young man who later became the father of the inventor.

In 1838, Alexander Melville, whose health was not good, was sent by his father to St John's, Newfoundland. During the four years he spent there, he seems not only to have recovered his health but to have developed in other ways. On his return to Britain he settled in London, where, following in his father's footsteps, he built up a successful practice in speech therapy in London. On a visit to Edinburgh he met and later married Eliza Grace Symonds, an accomplished pianist who had seriously impaired hearing: before marriage she painted miniatures for a living. They decided to make their home in the Scottish capital, where he built up a successful practice as an elocutionist. Their first son, Melville James, was born in 1845, one year before Alexander, but died from tuberculosis at 25. Their second son, the future inventor, was born on 3 March 1847 and christened Alexander; his middle name of Graham was added later. A third son, Edward, also died from tuberculosis, before he was 17. All three sons went to the Royal High School for their formal education, where only the self-confident and outgoing Melville distinguished himself.

Alexander, at the age of 15, was sent to London to live for a few months with his 70-year-old grandfather, by this time a widower, under whose influence he matured into a studious thoughtful young man. When his father came to take him back to Edinburgh, they called on Sir Charles Wheatstone, who is famous for his electrical work, including the discovery of telegraphy, but who was also interested in the physiology of vision and speech. Alexander took a school-teaching job at the old town of Elgin, on the Morayshire coast. He would have studied Helmholtz's famous book on the theory of sound but could not understand German; later he was able to read the book in a French translation. Independently, he began to try teaching deaf children to speak, with some success, and this was to become his main profession. He used a method developed by his father, called Visible Speech, a system for recording vocal sounds. Accompanied by his brother David, Melville went to the United States to lecture on the subject, where they were well-received. Meanwhile Alexander, back in London, looked after his father's affairs. After his brother Melville died in 1870, the older Bells and Alexander decided that their future lay on the other side of the Atlantic, and so they decided to settle in Canada just outside the industrial town of Brantford near Montreal and not far from the American border. Alexander secured some short-term appointments in Boston, teaching the deaf to speak. In Boston, he met a patent lawyer named Gardiner Hubbard, who had a deaf daughter Mabel; Bell started to teach her how to speak properly.

Before he left England, Bell had become interested in telegraphy and particularly the problem of sending speech by telegraph, instead of just coded messages. As a first step, he sought, like several others at this time, a method of sending several messages simultaneously along the same line. Unlike Edison, he favoured a harmonic approach, sending several pitches simultaneously and unscrambling them with tuned receivers. He was conscious that he was lacking in technical knowledge and skill to make a practical model. To obtain a patent he would need a specialist lawyer, which would be expensive, so he decided to rely on secrecy instead. To improve his scientific knowledge, he attended some free lectures on science at the Massachusetts Institute of Technology (MIT). He was offered and accepted a professorship of vocal physiology at Boston University, which brought him a small income and an office where he could see private pupils. He commuted by train from Salem, where the mother of a young boy provided him with free rooms and board in return for giving private tuition to her quiet and withdrawn son. Bell was already a night owl, doing his best work at night and snatching a few hours of sleep during the day, a habit he maintained to the end of his life.

Hubbard offered Bell funds in return for a share in patent rights. They discovered that a practical electrician named Elisha Gray, also of Boston, had already applied for a patent for at least one element of a harmonic system, and so a race started to construct a working model. Bell understood the theory of speech far better than Gray but the latter had more practical skills and was aiming at the transmission of music. Eventually, Bell was ready to demonstrate his apparatus and drew up an American patent application. He impressed Joseph Henry but William Orton, president of the Western Union Telegraph Company, which controlled a vast network of telegraph lines, was non-committal, partly because he had a personal dislike for Hubbard. Western Union made plans to compete with the Bell Telephone Company. Hubbard had a strong interest in telegraphy and was a fierce opponent of Western Union.

In 1874, Bell took 'first papers' for naturalization, because it was difficult for a non-American to obtain an American patent. He was still teaching Mabel Hubbard and the following year they became engaged. Although her parents approved of the match, they thought that, at 17, she was too young for marriage. Although Bell wished further to delay submitting the patent application that he had prepared, Hubbard was afraid that Gray would get in first and took it upon himself to file Bell's application. In fact, Gray, simultaneously filed what was called a caveat, which meant that there would have to be a formal hearing on Bell's application. Orton stood behind Gray, aiming to force Bell into buying Gray off, since until the caveat was withdrawn he would be unable to exploit his invention commercially without running the risk of protracted lawsuits. Although Bell was confident of winning, he felt it best to offer Western Union a share in his invention to clear the matter up. Orton refused an offer from Hubbard of all rights to the invention for $100,000, surely a bargain.

Meanwhile the demonstrations of what was now called the telephone became more and more convincing, although this did not translate into funding for further research. Bell was envious of the facilities of the laboratory at Menlo Park that were at Edison's disposal. Moreover, he was hoping for material rewards, which would enable him to get married to Mabel Hubbard. Early in 1876, her parents gave their consent and the wedding was held in May that year. Bell gave Mabel almost all of his shares in the telephone company that had been set up to market the invention. After a honeymoon at Niagara Falls, they set off to spend a year in Europe, where Bell needed to secure a British patent for his invention and to do something about other European rights. Also, he wanted to show his wife the places in Britain where he grew up and introduce her to his grandfather and other members of the Bell family.

Bell was no business man and made a lot of mistakes. Business letters were pouring in from all over the world from people who wanted to buy one of his telephones. However, the policy of the company was to lease, rather than sell, telephones. Much time and energy was spent in defending the patents. Once Bell had successfully demonstrated his invention, Edison also became interested in telephone research. With the superior facilities at his disposal, Edison was able to improve on Bell's design in several respects, especially in the design of the microphone. After 1877, Bell had little active connection with the telephone industry, but he continued with his experimental work, for example, in the design of aeroplanes.

The Bells decided to live in Washington. They began renting a house there in 1879, and three years later bought a grand mansion on Scott Circle. This was severely damaged by fire five years later, but rebuilt much as before. The Bells sold it in 1879, and had another house built to their own specifications on Dupont Circle, which was their home for the rest of their lives. Washington enjoys mild winters but very hot summers and so they also had a summer home built on Cape Breton, on the coast of Nova Scotia.

They also travelled a lot; he was an international celebrity and, when in Washington, the Bells entertained many of the good and the great at their home. He was also one of the early members of the nearby Cosmos Club, which, like its London counterpart, the Athenaeum, was a place where intellectuals, especially scientists, congregated. Their first child, a daughter, Elsie, was born in 1877, a second daughter, Marian, in 1880; after two premature babies, who did not survive for long, they had no more children. In 1882, Bell finally became a citizen of the United States.

Although teaching the deaf to speak remained his main purpose in life, Bell continued to have ideas for inventions, some of which he developed to the stage in which they might have become important. One was an early hydrofoil, which established a world speed record. He was also interested in heavier-than-air flight, like others of that time, but had no connection with the Wright brothers. However, he never repeated the success he had with the telephone. He used his wealth more to support others. He placed the journal *Science* on a sound footing and did valuable work for the Smithsonian, of which he was regent from 1898. In teaching the deaf to speak, he ran into strong opposition from those who believed that sign language was the way in which they should be taught to communicate. He was elected to the National Academy of Sciences and received many other honours, including the Hughes medal of the Royal Society, but the one he valued most was the Volta Prize, established by Bonaparte in honour of the Italian scientist Alessandro Volta for scientific achievement in electricity. Public schools for the deaf in Chicago and Cleveland bear his name, as does the American Association for the Deaf and Hard of Hearing. After a period of declining health, Alexander Graham Bell died at Cape Breton on 2 August 1922.

6 From Braun to Hertz

Ferdinand Braun (1850–1918)

Germany enjoyed a period of exceptional prosperity in the last quarter of the nineteenth century. The country overtook France in many ways, and engineering was one of these. Electrical engineering was a German speciality. Ferdinand Braun was born on 6 June 1850 at Fulda, a Catholic enclave in a Protestant region not far from Frankfurt. His father was a minor civil servant, who married the daughter of his superior. Ferdinand, their youngest son, had, altogether, four brothers and two sisters. After leaving the local gymnasium, Braun began studying physics at the minor University of Marburg but he soon moved to Berlin, where he received his doctorate in 1872. Like Heinrich Hertz later on, he became a protégé of Helmholtz. Two years later, as a young gymnasium instructor in Leipzig, he wrote his first book *Der Junger Mathematiker und Naturforscher*. He then progressed up the academic ladder, being außerordentliche professor first in Marburg and then in Strasbourg, then ordentliche professor first in Tübingen and then back to Strasbourg, where he remained for almost the whole of the rest of his career, during which time the city was in German hands.

Braun was the first to investigate the rectifier effect in semiconductor crystals, the phenomenon behind most solid-state electronics. In 1897, he invented, but refused to patent, the cathode ray oscilloscope, which became the basis for the television tube, computer terminals and many other electronic devices. At the time, however, it was considered that his most important work was in the field of wireless telegraphy. Puzzled by the limited range of Marconi's transmitter, Braun experimented in 1898 with a resonant antenna circuit that provided a more efficient transfer of energy. This powerful and practical design was soon competing successfully with those of Marconi and others. Braun was the most formidable rival of Marconi, whom he regarded as an amateur who took a long time to solve a problem that a professional scientist would find quite easy.

In 1901, under pressure from the Kaiser, competing German interests in the new technology were merged into the Telefunken company, which soon became internationally important. Braun played an important

part in its formation. He later supplemented his wireless transmitter with the development of a crystal detector, multiple tuning and a directional antenna. Not satisfied with mere laboratory devices, Braun actively promoted their development and commercial exploitation. He negotiated for capital, secured patents and supervised lengthy commercial trials.

Meanwhile Braun remained at Strasbourg despite calls from Leipzig and Berlin. In 1909, he shared the Nobel Prize for Physics with Marconi, in recognition of their contributions to wireless telegraphy. Telefunken had established a wireless station on Long Island. As soon as the First World War began, Britain cut the German transatlantic cable and attempted to shut down this American outpost by requiring Marconi's company to sue Telefunken for patent infringement. Although his health was failing, Braun went to the United States to help defend the action but felt that he could not risk crossing the Atlantic again to return home. He lived quietly in Brooklyn and died there on 20 April 1918. His wife Amalie, whom he married in 1885, had died in Germany the previous year. They had two children, Siegfried, in 1886, and Hildegard, two years later. Not only a researcher and academic entrepreneur, Braun was also a personality of diverse interests who wrote satirical verse in his youth and painted watercolours in his later years. One of his last publications was a successful book, *Physics for Women*, which he wrote while he was in Brooklyn.

Hertha Ayrton (1854–1923)

We now return to England, and the first of our two women engineers. Phoebe Sarah Marks, afterwards to become famous as Hertha Ayrton, was born on 28 April 1854 in Portsea. Her parents, Alice Theresa and Levi Marks, were then living in the old Sussex town of Petworth, but it seems that her mother returned to the home of her own parents to give birth to Hertha, her third out of eight children. Her father, the son of a Polish innkeeper, had fled to England to escape the Jewish persecutions under the tsarist régime. His health, probably undermined by these youthful experiences, was never good. He traded as a clockmaker and jeweller from a shop in Petworth, until his last year when he took out a license to hawk his wares as a pedlar. An unworldly man, he was unsuccessful in business and when he died in 1861 his family was left in poverty.

Thus, Sarah, as she was known as a child, was mainly brought up by her mother Alice, whose parents were also Polish refugees. Alice was a remarkable woman, who brought with her the strict and narrow Judaism of her Polish forebears. Sarah's education began at a dame school in Portsea, to which her widowed mother had moved. When her exceptional ability began to show itself, she was sent to continue her education at a private school in north-west London kept by her maternal aunt Miriam and her husband Alphonse Hartog, through whom she got to know some remarkably gifted cousins.

At school, Sarah learnt fluent French from her uncle Alphonse, amongst other subjects. After school hours she gave private tuition to earn some money to send home to help her impoverished mother, who was struggling to raise her other children. Sarah, who was short in stature, had penetrating grey-green eyes and raven-black hair. Like her father, she was good-looking, but took no trouble over her personal appearance. Fiercely independent, free-thinking and stubborn, on one occasion when she felt unjustly accused of some misdemeanour, Sarah went on hunger strike for several days. At the age of 16, she told everyone that in future she wished to be known as Hertha, the earth goddess in one of Swinburne's poems.

In London, Hertha's social circle began to widen. The most important of her friends was Barbara Leigh Smith (1827–91), a first cousin of Florence Nightingale, who was an outspoken feminist and a prominent figure in the movement for women's emancipation, a cause dear to Hertha's own heart, as we shall see later. After Barbara married a French-born physician named Eugene Bodichon, she was generally known as Madame.

Madame was one of the founders of Girton College, Cambridge. She encouraged Hertha to try for a college scholarship. Although Hertha was not awarded this, she was nevertheless admitted to the college in 1876, supported by a loan from Madame Bodichon and her friends. At Girton, Hertha was prominent in the Choral Society and helped to form the famous Girton Fire Brigade, which is still going strong. After her first term, she was taken ill and left Cambridge for the rest of the year to convalesce. At this time, women were only permitted by the university to sit the Tripos examination in college, rather than in the Senate House with the other candidates, and however well they did they could not proceed to take their degree. When the time came, Hertha did not do at all well; she was only fifteenth in the third class. The university never granted her a degree, but she remained attached to the college until the end of her life. Cambridge did not grant degrees to women until 1948, when it was the last British university to do so.

After leaving Cambridge, Hertha and one of her classmates rented a flat in London where they gave private tuition. She invented a device designed to divide a line into any number of equal parts. Based on an idea of one of her cousins, this was accorded a favourable reception by architects, artists and engineers. After two years of preparing candidates for university entrance, she decided that she wanted to study applied physics and took a course at the Finsbury Technical College under William Ayrton, a pioneer in physics education and electrical engineering. The son of a barrister, he had studied mathematics at University College, London and electricity at Glasgow under William Thomson. For some years he had worked abroad, in India and Japan. After he returned to London, he was appointed lecturer at several of the technical colleges, of which Finsbury was one.

Ayrton had recently lost his first wife. He married Hertha in 1885; they raised two children, one daughter from each of his two marriages. Thanks to a legacy from Madame, Hertha was able to employ a housekeeper, which enabled her to start assisting her husband with his experiments on the electric arc. The arc, which is produced when an electric current flows between two electrodes, had been discovered in 1820 by Humphry Davy, and by this time it was in regular use when a very bright source of light or very high temperature was required. Soon she was conducting her own experiments at home while her husband looked after their daughters, and these led her to make some important discoveries.

Hertha began to lecture on her research, at home and abroad, and wrote her only book *The Electric Arc*, published in 1902 (and still in print, see Ayrton, 2007). This was a practical treatment as well as theoretical; later it was partly superseded by technical developments in the practice of welding but for arcs between carbon electrodes it remained the standard work, containing the only complete history of the subject. She worked on the design of electric searchlights for the Royal Navy between 1904 and 1908, although her improvements were attributed to her husband. He had been elected a fellow of the Royal Society in 1881 and was awarded a Royal Medal 20 years later. Efforts were made to get Hertha elected to the Royal Society as well but, although she had strong support, the statutes in force at that time blocked the nomination of women, and the attempt was not repeated even after the statutes had been changed. The argument was that a woman's person was, in common law, covered by that of her father or husband; no woman was elected until 1945. However, her work was recognized in 1906 by the award of the Society's Hughes Medal for 'an original discovery in the physical sciences, especially in the field of electricity and magnetism or their applications'. The previous year, she had been admitted a full member of the Institution of Electrical Engineers.

Hertha gave an invited address in French on her research into the electric arc at the International Electric Congress of 1900, held in Paris. She returned to Paris in 1911 to lecture, also in French, to the Société de Physique on her theory concerning the formation of ripples in sand under standing waves in water. When the Curies came to London to report on their isolation of the element radium, Hertha found a kindred spirit in Marie Curie. Whenever she was in Paris, she would visit Mme Curie who, on her own in 1912 and with her daughters the following year, came to stay with her in England.

When poison gas was used in the First World War, Hertha succeeded, not without difficulty, in persuading the War Office to make use of a simple invention of hers, which could disperse clouds of gas when used in a certain way, based on her understanding of the reasons for the formation of sand-ripples. Over a hundred thousand of these Ayrton fans, which also had other applications, were used on the Western Front.

Being left-wing in her political views, Hertha joined the new Labour Party when it was formed. She had always been a staunch supporter of women's rights, as was her husband. In 1899, she presided over the science section of the second meeting of the International Congress of Women. She played a leading part in the suffragette movement, especially after the war, and was proud that her daughter became an even more militant suffragette than herself. In 1898, her mother died; her husband died ten years later. She herself died on 26 August 1923, at the age of 69.

Sir Charles Parsons (1854–1931)

Like Scotland, Ireland has produced a number of fine engineers, of whom Charles Parsons was the most remarkable. He was the youngest of the six sons of William, third Earl of Rosse, a distinguished astronomer who was president of the Royal Society 1848–54 and died in 1867. His mother Mary, née Field, daughter of a Yorkshire squire, excelled in every form of handicraft, including modelling and photography. Two of their sons died in early boyhood. The remaining four spent their boyhoods at the Irish family seat, Birr castle, Parsonstown, where they were brought up by their mother. Three of the four were fascinated by mechanics and engineering. They never went to school, but spent much of their time in their late

father's excellent workshop. Amongst other machines, they made a steam wagon with which they used to give rides to visitors. They were also keen on sailing, and eventually bought the late Robert Stephenson's capacious yacht *Titania*, which took them to other parts of the British Isles and to the Low Countries.

At 17, Charles followed family tradition by spending a year at Trinity College, Dublin, and then went up to St John's College, Cambridge, where he was prominent in the college boat club. At Cambridge, Parsons was described as a 'spare athletic young man, with bright blue eyes and fair reddish hair'. He had an eager and enthusiastic disposition, although he was constitutionally shy, self-contained and inexpressive. He was able to pursue the same problem almost endlessly. Perhaps he had some degree of autism. The people with whom he was most at ease socially were eccentric Irish aristocrats like himself.

Parsons graduated from Cambridge as 11th wrangler in 1877. Since there was no engineering department at the University, he went to the Armstrong Whitworth engineering works near Newcastle as a pupil-apprentice, where he impressed Sir William Armstrong with his industry and ability.

He was particularly interested in the development of the underwater propulsion of torpedoes by means of rockets. After completing his apprenticeship, he worked in the firm of Kitson and Company at Leeds, which allowed him to try out some of his ideas. He then became a junior partner in the firm of Clarke Chapman of Gateshead, who were making vacuum pumps for exhausting the new electric light bulbs produced by Swan's firm in Newcastle. Shortly before, he had become engaged to Katherine Bethel, of Rise Park in the East Riding of Yorkshire, and they were married in 1883. They honeymooned in America, exploring the continent from coast to coast with horse and buggy. In Pittsburgh, they went over the engineering works. On their return, their first child, Rachel Mary, was born. Their second child, Algernon George, was born less than two years later; he was killed in action towards the end of the First Word War, causing Parsons long-lasting grief.

Parsons realized the urgent need for a high-speed engine capable of driving dynamos directly. For this, a reciprocating engine is inherently unsuitable, and he was led to develop the steam turbine, the first patents for which he took out in 1884. The principle was not new, but unlike earlier inventors he realized that the drop in pressure must take place in stages. His first engine, built in 1884, developed 10 horsepower at 18,000 revolutions per minute. Later he introduced a condenser to utilize low-pressure steam formerly wasted, and the use of high pressure superheated steam. He used it to drive dynamos of a novel but reliable type which, although he was not an electrical engineer, he designed and constructed himself.

Parsons had assigned his patents to Clarke Chapman, as part of the partnership agreement, but regretted this when he found that although his fellow-partners were interested in developing turbines they did not agree with the need for haste. He therefore dissolved the partnership and, after a lawsuit, they agreed to sell the patents back to him. Parsons then established a successful business of his own, and this concentrated on developing the turbine further for use at sea; his knowledge of seamanship stood him in good stead. In 1897, his *Turbinia*, which was capable of 34 knots, created a sensation at the great naval review in the Solent marking Queen Victoria's diamond jubilee. He developed reduction gearing, which made it possible to use turbines in slow powerful vessels as well as light fast ones. In 1906, the British Admiralty installed them in the famous Dreadnought series of battleships; the great Cunarders *Lusitania* and *Mauretania* followed. By 1915, they had become the predominant form of propulsion on the high seas.

Parsons was a capable and successful businessman, but not good at management. Many very able and loyal senior employees found it impossible to work with him. Yet the British physicist J. J. Thompson wrote of working with him on the wartime Inventions Board:

> He was a very agreeable as well as a most efficient colleague. I have rarely been on a Board where there was less friction and where proceedings were more harmonious. He had the engineering instinct more fully developed than anyone I had ever met, beside being by far the greatest and most original engineer this country has had since the time of Watt, he was one of the kindest and most steadfast of friends.

Parsons did not believe in monopolizing the applications of his ideas but granted licenses to avoid ruinous legal disputes. He sold the American rights to his turbines to Westinghouse, who after making some improvements used them in the Niagara Falls power station. The profits from his manufacturing business went mainly into further development; he became rich from licensing fees and royalties. However, he was very reluctant to license the innovations of others, preferring to find another way of achieving the same end.

Invention and engineering were his recreation as well as his vocation; like Edison Parsons enjoyed his work. Altogether, he took out 300 patents. Turbines were his speciality but by no means his sole interest. For example, he was an early supporter of aeronautical engineering. He purchased an Irish optical factory and started to improve the methods of making optical glass, initially for use in searchlights but later for the large telescopes used in astronomy. He invented a novel form of sound amplifier but was too busy to develop it. His most costly failure was an attempt to make artificial diamonds by submitting carbon to high pressures and temperatures.

The importance of Parsons' work was widely recognized in his lifetime. He was made Knight Commander of the Order of the Bath in 1911, and admitted to the Order of Merit in the same year. He was elected to the Royal Society in 1898, and became a vice-president ten years later. He was awarded the Rumford medal and the Copley, as well as numerous honours from other bodies. It is recorded that although an excellent mathematician, formal calculations interested Parsons very little. He generally reached his results almost instinctively, by obscure mental processes, which he himself did not properly understand. He died at the age of 76 while on a cruise with his wife in the West Indies on 13 February 1931.

Granville Woods (1856–1910)

Charles Parsons was born rich and became richer. The subject of my next profile was born poor and remained poor. The son of Tailer and Martha Woods, Granville was born in Australia on 25 April 1856. Apparently, his mother's father was a Malay Indian and his other grandparents were full-blooded Australian aborigines, born in the wilds back of Melbourne. Although his complexion was coffee-coloured rather than black, he was regarded as a Negro throughout his life, but not by himself. A newspaper declared him to be the greatest coloured inventor in the history of the race, and the equal, if not superior, to any inventor in the country.

Woods' early years were spent in Columbus, Ohio (according to some accounts of his life he was born there). When he was 10 he left school and went to work in a machine shop. At 16, he moved from Ohio to Missouri and took a job as a fireman and engineer (driver) on a railroad. An avid reader, during his leisure time he borrowed books on electricity from the local library and from friends. Moving to Springfield, Illinois and then to New York City, Woods found work wherever he could, first in a steel mill and then in another machine shop, but he was determined to become a proper engineer. Where he did so is uncertain, but he attended courses at an electrical and mechanical engineering school. With his new knowledge, he secured a job as engineer on *Ironsides*, a British steamship. After working on this ship for two years, he took a more senior engineering job on another railroad, the Danville and Southeastern. Unfortunately, the company, being in financial trouble, paid its employees in scrip of little value, and so Woods had barely enough to live on.

Woods' first invention was an elevator (lift) signalling system using electromagnetic induction. He showed his plans to a patent attorney, who pointed out that elevators could be signalled by the ordinary method very cheaply. However, Woods knew that there was a pressing need for a means of communication between trains and railway stations, and so he adapted his inductive communication device for this purpose. Unfortunately, he did not have the funds to develop this idea. Moreover, he came down with a severe case of smallpox. This not only kept him bedridden for several months but also left him extremely weak for the next few years. After he regained his health, he began to work on other inventions, which he hoped would finance the development of his system of inductive communication. He tried to interest the Westinghouse Air Brake company in an idea he had for an electromagnetic braking system for trains.

In 1883, Woods filed a patent application for a steam-boiler furnace. He also filed applications for two telephonic devices. One, he called the synchronous multiplex railway telegraph, whereby the dispatcher could discover the position of any train at a glance. The system also provided for telegraphing to and from the train when it was in motion. The same lines could also be used for local messages without interference with regular train signals. In using the device, there was no possibility of collisions between trains as each train would always be informed of the position of any other while in motion. Woods sued the Thomas Edison Company for infringement and won his case. He often had to defend his patents against others.

The other was a device that combined the telegraph with the telephone. Woods called it telegraphony and patented it in 1885. Instead of reading or writing the Morse code signals, an operator could speak near the telegraph key. This made it possible to receive both oral and signal messages clearly over the same line without making changes in the instrument and without understanding the Morse code. Woods' patented telegraphony was purchased by the Bell Telephone Company for a hundred dollars, but they did not develop it.

In 1886, Woods and two Cincinnati businessmen formed the Woods Electric Company in Kentucky. Woods signed a working agreement with the company, expecting that he would then be able to develop some of his inventions, but unfortunately his partners preferred to make money by selling the patent rights to the inventions, not by developing them. Moreover, the company only paid Woods' salary occasionally and so to make ends meet he opened a small machinery repair shop in Cincinnati. At the same time he hunted for a way to extricate himself from the contract he had signed, and in 1890 he succeeded by forming a different Woods Electrical Company, incorporated in Ohio rather than Kentucky.

Woods was now free to apply for patents for inventions that the original company would not develop. In 1890, Woods moved to New York City and joined his brother, Lyates Woods, who was also an inventor, but the best he could do for himself was a poorly paid job as porter on the Manhattan Elevated Railway. There was also an ailing sister in Ohio whom he supported, although he could hardly afford to do so. In 1891, Woods was still looking for investors, and an advertisement by the American Patent Agency attracted his attention. 'Inventions promoted and companies formed for inventors' were declared to be its specialities. The manager, James Zerbe, showed interest in Woods' inventions, particularly his design for a simple electric railway. A partnership was formed, called the American Engineering Company, to which Woods was contracted as a consultant. He expected that Zerbe and his associates would help him develop his invention but it soon became clear that their real intention was to deprive him of it. This was not the only time Woods was cheated of his intellectual property.

Whereas his partners, especially Zerbe, deceived him, Woods was not entirely innocent. In 1891, he began secretly filing a patent application for his latest version of the electric railway system, based on work he had done before he became an employee of the company. The following year, he began to realize that he had been cheated and put out a warning in a newspaper to the effect that the company was offering for sale an electric street railway system covered by Woods' patents, using plans purloined from the inventor and without his consent. Zerbe promptly sued Woods for libel and had him held in jail for several days. When the libel action came to trial, Zerbe lost his case and his unsavoury history came to light.

By the end of 1893, Woods had been awarded 20 patents; afterwards he was to be awarded 25 more. He had learned from bitter experience how to use his inventive skills for financial gain, modest as it was; he was involved in almost constant litigation, and what he earned mainly went to pay lawyers. Woods faced as many defeats as victories. Legal fees he could barely afford and powerful enemies in business and politics made his life a struggle right to the end. He died after a stroke in 1910.

Some people considered the third rail to be his greatest invention. Used in subway systems throughout the world, the third rail put electrical conductors along the path of the train so that it would receive the current directly. Woods received a patent for this in 1901 and sold the invention to General Electric shortly after. Others believed that Woods' air brake technology was equally important. Starting in 1902, he had developed several devices that led to the automatic air brake, which he sold to the Westinghouse Company He was determined to make a living as an independent inventor, not as an employee, but he was always lacking in the necessary capital and connections.

Nikola Tesla (1856–1943)

Nikola Tesla was one of the most brilliant inventors of his age: the word genius does not seem inappropriate in his case. The inventor's forebears were Serbs living in Croatia, who (if they were men) entered the Church or the Army. His father, Milutin, was politically active, wrote poetry and entered the Orthodox priesthood. In 1847, Milutin married Djouka Mandić, daughter of one of the more prominent Serbian families. Although illiterate, she was cultured, intelligent and hardworking; also gifted with an unusually

retentive memory. Nikola recalled that his father spoke many languages fluently and also understood mathematics well. He was an omnivorous reader and possessed a large library, which his son made good use of. However Tesla ascribed his inventive powers more to his mother.

In his childhood, the Tesla family moved from the rural hamlet of Smiljan, where he was born at midnight 9–10 July 1856, to the town of Gospić, where Milutin was pastor and taught at the Real Gymnasium, which his son attended. After his brilliant elder brother, Dane, died from injuries sustained when he was thrown from his horse, Nikola moved to the city of Karlovac (Carlstadt), where there was a Higher Real Gymnasium with good teachers. Here, he apparently contracted malaria, which persisted through most of his later life, and this was followed by cholera, which was nearly fatal.

He persuaded his father that he wanted to become an engineer, not to make his career in the church. To prepare for this, he went to the polytechnic school in Graz, the capital of Austrian Styria, which had a high reputation for science. Although he studied so hard that his teachers worried about his health he also sowed some wild oats, gambling and womanizing. Without graduating from the polytechnic, he transferred for one semester to the University of Prague, where he studied informally under some of the leading thinkers of that time. After that he set out to earn his living.

His first job was as a draughtsman in the engineering department of the Central Telegraph Office of the government in Budapest. When a telephone exchange opened in the Hungarian capital, he moved to the more interesting job of checking the lines and repairing the equipment. He took the various instruments apart and designed ways of improving them. However, his main goal during this period was to design an electric motor that would harness alternating current without the cumbersome intermediaries that direct current required. Eventually, he saw that the solution was to use an out-of-phase pair of currents. Others were working on the same lines but when Tesla applied for a patent his priority was confirmed. Direct current electricity at normal voltage could only be transmitted for short distances, even when just required for lighting, but alternating current could be economically transmitted for hundreds of miles and, after Tesla, used to power industrial machinery.

In 1882, Tesla left Budapest for Paris, where the Edison Continental Company was manufacturing dynamos, motors and lighting systems under the Edison patents. He was employed as a junior engineer in the company's factory at Ivry-sur-Seine. When Edison lighting systems were installed, the generators supplied often broke down until Tesla invented automatic regulators. The firm sent him to Strasbourg where the system was being installed, and his job was to investigate any problems that arose. He discovered that the wiring was defective but while he was putting this right he was, on the side, constructing his first electric motor. When he demonstrated his motor to some of the leading citizens of Strasbourg, they were impressed but were unwilling to invest money to enable its commercial possibilities to be realized.

So, in the spring of 1884, Tesla went to New York, where he hoped to find investors who would help him manufacture his motors. He met Thomas Edison, the wizard of Menlo Park, who was impressed by the young man and employed him as a consultant, without agreeing the financial terms. After a while, when Tesla asked for payment for his services, Edison prevaricated, whereupon Tesla walked out, understandably feeling that he had been cheated. Working independently, Tesla began to take out American patents for his inventions and formed the Tesla Electric Light and Manufacturing Company to exploit them. Before long, his backers forced him out of the company and so Tesla founded another company, the Tesla Electric Company, to exploit his inventions for alternating current technology. Westinghouse, as we know, was also developing alternating current systems, based on different principles, and building power stations, which produced high-tension alternating current, stepped down at the point of use for safety's sake. However, Tesla's system was far better and Westinghouse decided to purchase the American rights to it. He offered the equivalent of $75,000 in instalments, plus royalties amounting to $180,000 and 200 shares in the Westinghouse Company. Tesla accepted and as part of the deal agreed to spend a year working at the Westinghouse facility in Pittsburgh.

It was a frustrating year, in many ways, and Tesla was glad to leave Pittsburgh at the end of it and return first to New York and then to Europe. There he heard about the discoveries of Hertz, on the propagation of electromagnetic waves. In Europe, he met some of the leading scientists of his day, and also found time to visit his family in Croatia. When he returned to Manhattan, he established a laboratory with equipment loaned by Westinghouse. He contacted Michael Pupin, a fellow Serbian, who had emigrated to the United States in 1874. They had much in common, and together made several profitable inventions, based on Tesla's principles.

In 1891, Tesla visited London, where he lectured to a distinguished audience at the Institution of Electrical Engineers, at which he demonstrated some of his more spectacular inventions with remarkable showmanship. He repeated this at the Royal Institution, and then went on to Paris, where he was equally successful. Hearing that his mother was dying, he went to Gospić to see her for the last time, after which he met a delegation of Serbian scientists, lectured at the University of Zagreb, and at an audience with the king was awarded the title of Grand Officer of the order of St Michael. He also went to Berlin, to pay his respects to Helmholtz, and to Bonn, to meet Hertz, whose experiments he had replicated. They did not get on well, but Hertz was already suffering from the disease that led to his untimely death.

Before returning to New York, Tesla had been elected a fellow of the Royal Society. The lectures he gave in various places drew large audiences and he was lionized in the popular press. He gave demonstrations at the Chicago World's Fair of 1893. In these demonstrations he passed high frequency electricity through his body without, apparently, feeling anything or suffering any ill-effects. The same year when contracts were awarded for harnessing Niagara Falls technology based on Tesla's polyphase system was adopted.

In 1895, there was a fire in Tesla's laboratory, originating in the dry-cleaning establishment on the floor below, which not only destroyed much of his equipment but also all his notes and papers. Fortunately, his photographic memory helped him recover some of the written material, which would otherwise have been lost. Since he was not insured, he suffered financially when Westinghouse started charging him for machinery he had borrowed from his company. Edison gave Tesla temporary use of facilities at Menlo Park while he was arranging a new laboratory for himself in Greenwich Village.

Meanwhile, the Niagara scheme was successfully completed and Tesla received much of the credit for this. At this stage, he was interested in lighting and he invented both gaseous tube lighting and fluorescent lighting, although these were not commercially exploited until much later. He also found Roentgen rays, which had just been discovered, and radio waves. In the case of the latter, he was in a race with Marconi, with whom Edison was about to join forces.

In 1897, the luxurious Waldorf Astoria, the tallest hotel in the world, opened its doors and Tesla moved in and made it his base for the next 20 years. In 1899, he went out to Colorado where he constructed an experimental wireless station, using the El Paso electrical station as its source of power. Tesla was fascinated by a stupendous electrical storm he observed over the nearby Rocky Mountains. Tesla's imagination ran away with him when he found that he was receiving radio signals from a mysterious source, which he thought must be one of the planets. It seems likely that he was picking up signals from Marconi's trials.

The next year, Tesla returned to New York, without paying the outstanding account with the local power company. With the backing of J Pierpont Morgan, he purchased a 200-acre tract at Wardencliffe on Long Island sound, some 65 miles from Manhattan. There he planned to construct a wireless transmission station from which he could communicate with almost anywhere in the world. Marconi, at the same time, had been secretly transmitting messages between England and Newfoundland, using long waves that followed the curvature of the earth. The plan included a tower, originally to be 600 feet in height, had been scaled down to 200 feet when it was erected in 1892. At Wardencliffe, there was also extensive tunnelling underground, for some mysterious reason. However, creditors were now hounding Tesla and began to remove some of the heavier equipment, which belonged to the Westinghouse Company. The tower was broken into by vandals, who wrecked valuable equipment, and there were rumours that it was going to be destroyed by dynamite. In fact, it was demolished by a salvage company for the materials it contained.

By 1906, Tesla was in serious financial trouble on all sides. He continued to throw off ideas for inventions, such as a speedometer for motor cars, but they were not followed up. If he had put aside his more grandiose visions for a time in order to bring some of these into profitable enterprises, he might have been able to continue. Often he did not take out patents or defend them when he did; as a result, he was frequently the victim of piracy. He suffered a nervous collapse. His commercial sponsors began to desert him, since he habitually used money they provided for development to fund further research.

Tesla had become a United States citizen in 1891. A rumour that Edison and Tesla were to share the Nobel Prize in Physics for 1915 turned out to be unfounded, but Tesla was awarded the Edison Medal of the American Institute of Electrical Engineers in 1917. Twenty years later his 81st birthday was celebrated with the award of the Order of the White Lion from Czechoslovakia, and the Grand Cordon of the White Eagle from Yugoslavia. More helpfully, Belgrade also awarded him a pension of $600 a month. In his declining years, Tesla lived alone and became more and more eccentric. He adored pigeons and devoted much of his time to feeding and caring for them. He was 86 years old when he died in New York City on 7 January 1943.

Tesla had the ability to visualize in great detail. He claimed that he could see in his mind the structure of a machine, put it in motion, detect problems, make adjustments and thus design a perfect invention without ever having placed pen to paper or having performed an experiment. These mental characteristics brought him extraordinary insights. He experienced vivid unsolicited images, sometimes accompanied by strong flashes of light. To protect his secrets, he did not commit his major inventions to paper but depended on an almost infallible memory for their preservation.

Tesla had many characteristics of Asperger's syndrome. He was a visual thinker who had an eidetic memory. He hated touching another person's body and had a particular aversion to pearls. Except for one long-standing platonic relationship with a married woman he did not enjoy female company. He was always well dressed, in a formal, old-fashioned style. He preferred to buy new handkerchiefs or gloves rather than have the old ones washed or cleaned. He insisted that his table in the dining room of his hotel was never used by anyone else. He was punctual to a fault, usually working through the night. He was a generous but demanding employer. After his death, his reputation declined at first but recovered as some of his more speculative ideas were successfully developed. On the centenary of his birth in 1956 an international electric unit – the magnetic flux density – was named the Tesla.

Heinrich Hertz (1857–1894)

The name of Heinrich Hertz, the discoverer of radio waves, has already been mentioned several times. He was born in the free Hanseatic city of Hamburg on 22 February 1857, and grew up in a prosperous and cultured family. His father Gustav was a barrister and later a judge of appeal, with a seat in the Senate, his mother was Anna Elizabeth (née Pfefferkorn). He had three younger brothers and one younger sister. Hertz was a Lutheran, although his father's family were Jewish. At age six, Hertz entered a strict private school where his 'benign and understanding' mother watched over his progress: he was always first in his class. He had an uncommon gift for languages both modern and ancient. He left the private school at 15 to enter the Johanneum Gymnasium, where he was first of his class in Greek; at the same time he took private lessons in Arabic. Very early on, Hertz showed a practical bent; at the age of 12, he had a workbench and woodworking tools. Later, he acquired a lathe and with it made spectral and other physical apparatus. On Sundays, he attended trade school for lessons

in mechanical drawing. His skill in sketching and painting marked the limit of his artistic talent; he was totally unmusical.

After taking his Abitur in 1875, Hertz went to Frankfurt to prepare for a career in structural engineering. He gained some work experience with a civil engineering firm in Frankfurt. After a brief spell, in 1876, in the Dresden Polytechnic, he put in his year of military service, serving with the railway regiment in Berlin. He then moved to Munich in 1877, with the intention of studying further at the technical university there. Since his schooldays, Hertz had been undecided between natural science and engineering. While preparing for engineering, he had regularly studied mathematics and natural science on the side. With his father's approval and promise of continued financial support, he chose to enrol at the University of Munich rather than the Technical University. The latter institution had a more vocational emphasis but its facilities included a good physics laboratory. The university by contrast promised a life of study and research, one that suited Hertz's scholarly, idealistic tastes. He was relieved at having decided on an academic and scientific career after long vacillation and was confident that he had decided rightly.

Hertz spent his first semester at the University of Munich studying mathematics. Following the advice of the experimental physicist Philipp von Jolly he read the works of the French masters, learning mathematics and mechanics in their historical development and deepening his identification with investigators of the past. Elliptic functions and the other parts of the newer mathematics he found too abstract, unlikely to be any practical use. Although Hertz thought that, when properly grasped, everything in nature is mathematical; throughout his career he was interested primarily in physical and only indirectly in mathematical problems. It was expected at this time that an intending physicist should have a grounding in experimental practice as well as in mathematics. Accordingly, Hertz spent his second semester at Munich in Jolly's laboratory in the university and his third at the Technical University. He found this practical experience immensely satisfying, especially after intensive mathematical studies.

After a year in Munich, Hertz was eager to make the customary student migration. After consultation, he decided in favour of Berlin, where he was attracted by the fame of Helmholtz and Kirchhoff. He attended Kirchhoff's lectures on theoretical physics but found little new in them. Helmholtz was a poor lecturer, who simply read out verbatim passages from the books he had written in a halting and ponderous voice. However, the great man took a special interest in the new arrival. The research environment in Berlin was highly competitive, especially in physics. Hertz found that a prize was being offered by the Berlin philosophical faculty for the solution of an experimental problem concerning electrical inertia. Although only in his second year of university study, he was anxious to begin original research and try for the prize. Helmholtz, who had proposed the problem in question and had great interest in its solution, provided Hertz with a room in his new Physics Institute, directed him to the relevant literature, and paid daily attention to his progress.

Helmholtz saw the two sides of physics as complementary, so that an experimentalist needed to have a good grasp of theory, and vice versa. Hertz showed himself to be an extremely persistent and self-disciplined researcher. Outside the laboratory Hertz wrote home that his greatest satisfaction lay in seeking and communicating new truths about nature. His belief in the conformity of the laws of nature to the laws of human logic was so strong that to discover a case of non-conformity would make him highly uncomfortable; he would spend hours closed off from the world pursuing the disagreement until he found the error. He won the prize in 1879, earning a medal, a first publication in *Annalen der Physik* in 1880, and the deepening respect of Helmholtz.

Helmholtz encouraged Hertz to compete for another, much more prestigious and valuable, prize, which was being offered by the Berlin Academy. The subject was to test experimentally the critical assumptions of Maxwell's theory of electromagnetism. Hertz decided against doing so, feeling that the project might take him three years to complete and that the outcome was uncertain. Instead, he wrote a doctoral dissertation on electromagnetic induction in rotating conductors, a purely theoretical work that took him only three months. It was a thorough study of a problem that had been partially treated by many others, from Arago and Faraday to Emil Jochmann and Maxwell. He submitted his dissertation in January 1880 and took his doctoral examination the following month, earning a magna cum laude, a distinction rarely awarded in Berlin. The same year Hertz began as a salaried assistant to Helmholtz in the practical work of the Berlin Physics Institute, a position he held for three years. He got to know Helmholtz and his second wife Anna well; they entertained in great style, bringing together intellectuals, scientists, artists and leaders of both government and industry. His duties left him time to write 15 research papers and with them to begin establishing a reputation. This work in his Berlin period is difficult to summarize because of its diversity, but it was mainly concerned with electricity.

Hertz was very happy to be working under Germany's greatest physicist and enjoying the use of the country's finest research facilities. When the Berlin Physical Society began meeting in the Physics Institute, he attended regularly, enjoying the sense of being at the centre of German physics. However, it was time for him to advance to a regular faculty appointment, and for this the first step was to serve as a privatdozent. It was at this time that mathematical physics began to be recognized as a separate subdiscipline in Germany, and Hertz's opportunity came when the University of Kiel requested a privatdozent in that field and Kirchhoff recommended Hertz for the post. So, in 1883, Hertz moved to Kiel, where he proved to be a success as a lecturer; by the second semester he drew an audience of 50 students, an impressive number for a small university. The drawback was that Kiel lacked a physics laboratory, so that he could not carry out much experimental work. Instead he returned to theoretical physics and wrote a penetrating study of Maxwell's theory of electromagnetism. When Kiel offered to promote him to the rank of extraordinarius he declined because he did not want to be a purely theoretical physicist. When this became known, the Technical University of Karlsruhe offered him the position of ordinarius in physics and once he had inspected the physical laboratory there he accepted.

Hertz spent four fruitful years at Karlsruhe, from 1885 to 1889. His stay began inauspiciously: for a time, he was lonely and uncertain about what research to undertake next. In July 1886, after a three-month courtship, he married Elizabeth Doll, the daughter of a colleague. A few months later he began the experimental studies that were to make him world-famous. He settled the problem proposed for the Berlin Academy prize by confirming the existence of the electromagnetic radiation predicted by Maxwell, which extended the visible spectrum of light into what we now call the radio spectrum. He not only showed that electromagnetic waves exist but that they can be propagated in free space. The nine papers Hertz published on these researches at Karlsruhe won him immediate international recognition. With his experiments, Hertz had gone far towards his goal of testing Maxwell's theory decisively.

Helmholtz informed the Berlin Physical Society of Hertz's demonstration of the existence of these 'electric waves' in these words: 'Gentlemen, I have to communicate to you today the most important discovery of the century.' In summing up the significance of Hertz's experiments, Helmholtz said that they showed that light and electricity are very closely connected. Hertz was asked to lecture and repeat his experiments in Berlin and elsewhere. He took no interest in the commercial applications made possible by his revolutionary discovery, which had to wait for the development of wireless telegraphy by Braun, Marconi and others after his death.

In September 1888, the University of Giessen tried to recruit Hertz, while the Ministry of Culture pressed him to consider Berlin instead. At 31 he considered that he was too young for a major position in German physics, which would draw him away from research. In any case Berlin wanted a mathematical physicist. In America, Clark University tried to recruit him to join Albert Michelson at its new physical institute, planned to be as splendid as Berlin's, and then Graz tried to secure him as Ludwig Boltzmann's successor; without success. Finally, when he was offered the physics chair at the Rheinish University of Bonn he accepted, more because of Bonn's attractive location than its scientific reputation.

So Hertz moved to Bonn in the spring of 1889, succeeding the great thermodynamicist Rudolf Clausius. He and his wife moved into the house that Clausius had lived in for 15 years; the historical connection mattered a lot to him. He found the Bonn Physical Institute cramped and the apparatus a jumble, which he had to put in order. Hermann Minkowski, then a privatdozent in mathematics, joined him in research into electromagnetism. He also had his one and only assistant in Philipp Lenard, later to win a Nobel Prize for Physics but notorious in Hitler's Germany as a fervent Nazi. According to Max Born, Lenard, at a conference in 1920, maintained that it was the theoretical works of Heinrich Hertz that were the product of his Jewish inheritance, not his more important experimental work. Hertz's research on electromagnetism not only facilitated the acceptance of Maxwell's theory among German physicists, it taught them the value of a good theoretical basis for experimental work. Hertz himself, once he had finished arranging his laboratory at Bonn, returned to experimental research, but became discouraged by repeated failures. He published two more classic papers on Maxwell's theory in the *Annalen*; and then turned to something quite different.

The principle of least action has a long history in mechanics and physics. Helmholtz had been studying it afresh and Hertz decided to follow his example by writing a purely theoretical study of the principles of mechanics. While Helmholtz and others looked on his work in this field with respect they suspended judgement, and Hertz' theories have never been generally accepted, although Ludwig Wittgenstein believed in them. Meanwhile Hertz was starting to suffer from ill health. The decline began at Karlsruhe with toothaches; all his teeth were extracted. Then his nose and throat became so painful that he had to stop work. The cause was diagnosed as a malignant bone condition that his physicians could not deal with. He had several surgical operations but to no avail and he died from blood poisoning on New Year's Day 1894 at the age of 36, almost the same lifespan as Mozart. His wife and two daughters emigrated to England from Nazi Germany in 1937 and settled in Cambridge.

After his early death, his mother wrote a most interesting account of his childhood. Hertz himself kept a diary, in which he recorded the events of his everyday life. With other material, including a short biography by Max von Laue, these memoirs have been published in English (Hertz, 1977). Appleyard (1930) concludes his admirable outline of Hertz' life and work by saying:

> Among those who knew him best, the remembrance that remains of him is of a man of amiable disposition, social, genial, a good lecturer, possessed of singular modesty, who gave himself no airs as of a great professor, and who, when speaking of his own discoveries, never

mentioned himself. When the Royal Society presented him with the Rumford medal, he silently disappeared for a few days – none knew why – and he returned as silently. The habit he formed early in his life of solving difficulties for himself continued with him; he preferred on occasion to puzzle things out in loneliness in the laboratory. His decision to follow pure science instead of a technical career was faithfully kept, and yet the importance of the part he played in the ultimate technical advance in electrical science is important beyond measure.

7 From Diesel to Marconi

RUDOLF DIESEL (1858–1913)

The internal combustion engine was developed in stages during the last quarter of the nineteenth century. The Otto four-stroke gas engine was introduced in 1876, the small revolutionary petrol engines of Gottlieb Daimler in the decade 1880–90. This was followed by the development of the diesel engine. Nowadays, these powerful, economical and reliable engines are to be found everywhere, not only for transport in trucks, locomotives and ships, but in many other situations. While diesel technology has been greatly developed over the years, the basic principles were established by the man whose life is profiled next.

Rudolf Christian Karl Diesel was born on 18 March 1858 of South German protestant parents in Paris, where he received his early education. His father, Theodor, was a craftsman in leather with an interest in spiritualism who, around 1850, had moved to the French capital, where he struggled to support his family by practising Mesmerism as well as his craft. His wife Elise, née Strobel, a native of Nuremberg who was living in London at the time of her marriage, was the more practical one. They had three children, Louise in 1856, who died in her teens, Rudolf in 1858 and Emma in 1860.

Their son Rudolf was fluent in French, gave private lessons in the language to supplement his income, and later spoke it at home with his family; his mother also taught him English. After the battle of Sedan in 1870, Germans who lived in France had to leave; Diesel and his parents went to England, where they experienced poverty. After a very short time Diesel was sent to stay with an uncle in Augsburg. He attended the industrial school there, where his work was considered outstanding. In 1875, he went on to the celebrated Technische Hochschule in Munich, where he studied thermodynamics under Carl von Linde, from whom he received the scientific groundwork on which he was to base his great invention.

After Diesel graduated from the Hochschule, he went to work at the Sulzer Brothers Machine Works in Winterthur to gain some practical experience of engineering. In 1880, he went back to Paris to work for a flourishing firm, owned by Linde, which was manufacturing refrigeration

plant. After only a year there, he was promoted to factory director. Although this kept him very busy, he found time to embark on his own career as an inventor. His aim was to design a heat engine that would be more efficient than the steam engines which were in general use. Carnot had shown that this was theoretically possible and Diesel was not the only one working on the problem. He constructed an expansion engine based on ammonia, making use of Linde's refrigeration technology; it was essentially a steam engine using ammonia in place of water.

When there proved to be too many technical problems associated with this, Diesel began instead to consider designing a compression engine, in which air is compressed at high pressure and during this process becomes hot enough to ignite fuel, which is sprayed into the cylinder at the end of the compression stroke. This is the operating principle of diesel engines, which have high thermal efficiency. In 1892, he obtained the German patent for this process and wrote a book entitled *The Theory and Construction of a Rational Heat Engine*. This aroused widespread interest when it was published in 1893 and he was able to persuade two leading engineering firms – Maschinenfabrik of Augsburg and Friedrich Krupp – to support him in its development, which took far longer than he anticipated.

Meanwhile, Diesel had been looking for a wife, and in 1883 he married Martha Flasche, the daughter of a notary from Remscheid and governess of the children of a German merchant living in Paris. They continued to live in the French capital, where they had three children in the next few years, Rudolf Junior in 1884, Hedy in 1885 and Eugen in 1889. In 1890, Diesel moved to a new post in Berlin where at first he was employed servicing Linde's refrigeration equipment. However, from 1893 he was able to spend most of his time perfecting his engine. He looked back on those years as a period of exceptional happiness. In 1897, he was satisfied with what he had achieved and an independent test confirmed the engine's thermal efficiency, which came close to the ideal of the Carnot cycle. However, his health was causing trouble. He was a workaholic, tormented by insomnia and blinding headaches. He may have suffered from manic depression. At any rate he needed to spend some time in a sanatorium.

In the mistaken belief that development had reached the stage when the engine was marketable, it was displayed in an important international exhibition at Munich in 1898, when interest in it was worldwide. He found investors in a company he formed to exploit his invention. The following year, production was begun at Augsburg but the product was of poor quality and there were frequent breakdowns. The engine was beset with teething troubles, and was being marketed before they had been rectified. At this critical period, Diesel was being treated in the sanatorium and so unable to put matters right. As a result he lost control of his company, although until 1905 he remained one of its directors. Most of the subsequent development work was done by others at the engine works of Maschinenfabrik in Augsburg, of Nobel in St Petersburg, and in France.

During this period, Diesel found himself under attack by rivals. He was caught up in legal battles to try and invalidate his patent. It could be shown that the engine as developed was not the same as the one described in the original patent application. It was also claimed that his invention was lacking in novelty. Already in 1824, Carnot had suggested the use of air as working medium and the use of compression ignition, and the advantages of high pressures and temperatures in a working cylinder were all well-known by the late 1880s. These disputes caused Diesel a great deal of trouble. In addition, he had a capacity for alienating people who were trying to be helpful.

In the 1890s Diesel became on paper one of Germany's richest men. In 1895, he moved with his family to Munich, conveniently near to Augsburg, and commissioned a grandiose villa in a fashionable area of the city.

This cost so much that it started to undermine his financial position. He and his wife had a busy social life, as part of the haute bourgeoisie of Munich, and travelled extensively. Diesel twice went to America on business. He was given a warm welcome but his impressions of American society were largely negative. He met Edison, who demonstrated his latest phonograph.

Diesel had long been interested in social questions, such as the social ills of the Industrial Revolution, the condition of the working class, and the growth of class conflict in late-nineteenth-century Europe. He wrote a book entitled *Solidarismus* (Diesel, 2007) expounding his socialistic ideas on the subject, similar to those of Saint-Simon and other French utopians, but when this was published in 1903, he found that it attracted little interest, although he regarded it as more important than his invention of the engine. However, he had not given up on engineering entirely, although there was nothing revolutionary in his work. He collaborated with others on the design of diesel engine railway locomotives and of sufficiently small engines to be used in road transport.

Diesel was not an astute businessman. By 1913, his financial position was deteriorating rapidly and he was struggling to keep up appearances. He had made some unwise investments in Galician oilfields and in real estate, and was faced with bankruptcy. In mid-September, when other members of his family happened to be absent, he burnt various papers related to his financial and business dealings. Diesel died on 29 or 30 September 1913 by falling overboard from the Antwerp-Harwich steamer Dresden. It seems almost certain that he took his own life. A full biography of Diesel has been written by his son, Eugen, and several other books about him depend on this as their main source for the facts of his life.

Elmer A Sperry (1860–1930)

Like Eads and Edison, Elmer Sperry was a native-born American. He was born on 12 October 1860 in the rural village of Cortland, New York State, the son of Stephen Sperry, farmer and carpenter, and Mary Burst, who died giving birth to him. He had a normal school education, which gave only an introduction to science. Although he also spent a year hearing lectures at Cornell University, his interest in machinery was chiefly aroused by a visit to the Centennial Exhibition at Philadelphia in 1876. From then on, he learned all he could about mechanical and electrical engineering, and was able to improve on the dynamos and arc lamps then being manufactured. With financial support from the Cortland Wagon Company, he moved to

Chicago where, on his 20th birthday, he opened his first factory for the manufacture of a complete arc-lighting system that he had developed. Soon his company installed arc-lights throughout the Midwest, although not in the city itself.

From 1887, Sperry made a stream of inventions, which required both mechanical and electrical insight. For his most promising patents his policy was to found his own firms to develop his ideas and exploit them commercially. Usually, however, he sold the rights and used the proceeds to finance further invention, often in a completely different sphere. Thus, he started in 1888 with electrical machinery for coal-mining; two years later it was tramcars, equipped with electric braking, power transmission and speed control. This led him into the emerging automobile market. He fought hard to acquire George B. Selden's broad automobile patent of 1879, based on internal combustion, but was defeated. Instead, he focused on electric cars, and produced some prototypes, but in 1900 sold his patents and left the field to others. The following decade was devoted to chemical engineering;

next he turned to compound diesel engines; and then the production of high intensity arc-lights. In each case, he made significant improvements in existing technology.

However, Sperry's name is remembered chiefly for the practical development of the gyroscope, whose properties he began to investigate about 1896. In 1908, he had devised a way to prevent ships rolling, which was adopted by the United States Navy to provide the stability necessary for long-range gunnery. There was an obvious need to replace magnetic compasses with compasses that were capable of functioning within iron hulls. Although a gyroscopic compass had already been developed in Germany, Sperry began competing with his own superior design, first used tentatively in 1910, in the *USS Delaware*, and soon became standard equipment. His gyroscopic stabilizer for ships was introduced in 1913. Other naval applications followed, and his equipment was fitted to warships throughout the world. Not everything he invented turned into a viable commercial proposition. The compound diesel engine, while technically feasible, turned out to have an excessive production cost after a great deal of money had been spent on it. Altogether Sperry held more than 400 patents and founded eight manufacturing companies, one of which evolved into the technological giant Sperry Rand Corporation.

At the age of 27, Sperry had married Zula Goodman, by whom he had four children. In the latter part of his life, they lived in Brooklyn, where he died on 16 June 1930. Of the children, it was his eldest son Lawrence who carried on his father's work, especially in the field of aeronautics. In 1914, his automatic aircraft stabilizer was awarded the first prize in the French aero club for safety devices, which he had designed with other members of the Sperry family. He also developed automatic pilots for aircraft and an aerial torpedo, the flying bomb, which was resurrected by the Germans in the Second World War, as we shall see. Lawrence, a pioneer aviator, was killed in a plane crash in 1923.

Sperry was a founding member of the American Institute of Electrical Engineers, sometime president of the American Society of Mechanical Engineers, and the recipient of many other honours. He was one of the few engineers to be elected to the National Academy of Sciences. Politically, he was very conservative, like so many American industrialists of that period, and a strong supporter of Herbert Hoover. He was fascinated by Japanese culture and respected deeply the Imperial Navy's dedication to technological progress. After the First World War, he fought hard to maintain good relations between the United States and Japan.

Wilbur Wright (1867–1912) and Orville Wright (1871–1948)

Like Edison, but with more justification, the sturdy and independent Wright brothers are regarded as an example of a distinctively American kind of engineer. Wilbur Wright, son of Milton and Susan Catherine (née Koerner), was born in the manufacturing town of Dayton, Ohio on 16 April 1867. His brother Orville was born in the same place on 18 August 1871. Their father, a kindly, broad-minded, cultivated man, was at that time minister of the small non-conformist Church of the United Brethren of Christ and editor of *The Religious Telescope*, its official organ; he later became a bishop in that church. As the family moved around in accordance with his pastoral work, the brothers received their early education at various public schools in the mid-West, but at home they took advantage of his good library. Neither went to college. While attending high school in Dayton, Orville opened a printing office and began publication of a small weekly newspaper called the *West Side News*. It was printed on a home-made press built mostly of wood and the odd bits of carriage iron to be found in those days about a barn. A year later, they published a four-page five-column daily known as the *Evening Item*. This publication, too large an undertaking for young men of their resources, lasted only a few months. Late in the year 1890, they printed *The Tattler*, a four-page weekly edited and almost entirely written, by Lawrence Dunbar, a black youth who later won fame as a poet.

When the cycling craze began to sweep America in 1892, the Wright brothers opened a shop for selling and repairing bicycles. Later they manufactured these and other small machines, such as typewriters. The bicycle business being of a seasonal nature, they had time in the winter which could be devoted to reading and other pursuits in which they were interested. As children some years before, they had built small toy helicopters and later had read with more than ordinary interest articles appearing in magazines on the subject of human flight.

After reading all they could find about previous experiments with models and with full-sized gliders and power-driven aeroplanes they became convinced of the need to devise a better system of control. The German pioneer, Otto Lilienthal, for example, had manipulated the weight of his body as a cyclist does. In the summer of 1899, they built a large box-kite to try out some of their ideas for maintaining equilibrium in flight and introduced some features, such as wing flaps, which are used in every aeroplane today. Octave Chanute, an engineer of Chicago, who had made many experiments in gliding, was impressed by what the brothers were doing and gave them some good advice. He may have introduced them to Cayley's work, which the brothers greatly respected. Later, he gave them further encouragement and offered financial support, which they declined.

Deciding that their ideas could only be tested properly in the air, they next built a two-winged glider able to carry a man, and tested it on the desolate sand dunes near Kitty Hawk, North Carolina, in the autumn of 1900. They were puzzled to find that although it was more stable than previous gliders, and could be controlled mechanically, its lifting power was much less than it should have been, according to their calculations, which had been based on the generally accepted tables of air pressures. Thinking that the disappointing results might be due to design faults of their own, they built a second glider adopting the wing curvature used by Lilienthal. The performance of this machine agreed with calculations no better than did the first. Evidently the fault must lie in the tables.

The Wrights then set up a small wind tunnel in their bicycle shop at Dayton in an attempt to get more accurate data for the calculation of aerodynamic phenomena. When the demands of their cycle business allowed, they designed new instruments and measured pressures on hundreds of aerofoils. They also determined the centre of pressure on cambered surfaces at the different angles of attack. They verified the extremely important conclusion, reached in their gliding experiments, that when the angle is decreased the centre of pressure moves backward rather than forward, as was the received wisdom. Although neither of them had received any training in science or engineering, their experimental work was of the highest order.

In 1902, after four years of experimental work, they were ready to test another glider. In over a thousand flights at Kitty Hawk it performed according to calculation, indicating that their wind tunnel data could be trusted. Before leaving Kitty Hawk, the brothers had already begun designs for a power-driven machine, not a glider, which would carry a man, with wings of such efficiency that they would not need an extremely light engine or an excessively frail structure.

In their gliding experiments of the past several years, they had perfected their system of control, and they took out a patent on this. A new problem now presented itself. Practically nothing regarding aircraft propellers working in the air could be learned from the literature. Marine propellers were designed from empirical formulae not applicable to air. The Wrights could not afford the money required to make extensive experiments and so were forced to work out for themselves a theory on which to base the design of propellers for their new machine. Using this theory, in conjunction with the knowledge of air pressures gained in their wind tunnel, they calculated in advance the performance to be expected of their propellers, both when standing and in flight. The thrust of these first propellers when actually measured was found to disagree less than two per cent from calculation.

The engine also posed problems. Those available commercially were too heavy, so they had to make their own. However, by the end of 1903, all these difficulties had been overcome. The machine, like all early aircraft, was a biplane, with the pilot sitting in front of the propellers. This aeroplane was tested at Kitty Hawk on 17 December 1903. Four flights were made with Wilbur as pilot. The last was the first time in history that a power-driven aeroplane had lifted a man into the air in free flight. A gust of wind blew the machine over, after it had landed, and major repairs were necessary before it could be flown again. Few were there to witness the historic event; the American press hardly noticed it at all.

The Wrights continued their experiments near Dayton in 1904 and 1905 for the purpose of perfecting their aeroplane. Tail spin was sometimes a problem; they dealt with it by making the rudder movable rather than fixed. They thought that the practical value of aircraft lay in delivering mail or in observing military operations. In the latter capacity, they attempted to interest a sceptical United States War Department in their invention. Failing in this, they began negotiations with foreign governments, and in 1907 they both went to Europe, taking one of their planes with them to give exhibition flights. Observers from Britain, France, Germany and Italy were greatly impressed by the way the machine easily became airborne and made sharp turns under perfect control.

Slowly, America began to realize what the brothers had achieved. Eventually, the United States War Department began to wake up to the seriousness of the situation and renewed negotiations with them in Europe. A contract was soon entered into with the United States government for the purchase of a plane, and Orville Wright, in September 1908, began test flights preparatory to delivery. On 9 September, at Fort Myer, near Washington, he made a flight of over one hour's duration. Later the plane, carrying a passenger, crashed and the pilot Orville Wright was seriously injured while the passenger was killed. Accidents, often fatal, were all too common in those days.

While convalescing from his injuries, Orville joined his brother Wilbur in France, where the latter was training pilots for a company that had acquired the Wright patents in France. A company was organized to exploit the Wright patents in the United States. Wilbur, the more inventive of the two brothers, died of typhoid in 1912. They had always worked together; it was their practice that when one advanced an idea the other would, on principle, try to find fault with it, so that time-wasting and costly mistakes were avoided. Orville, who was more of a businessman, worked for the company for a few years but then disposed of his interest in the patents; he was also kept busy receiving the many honours which were showered on them, until he died in 1948. He tried in vain to find somewhere in the United States where the aeroplane that made history could be put on display. Just before their historic flight, a flying machine designed by their rival S. P. Langley, the Secretary of the Smithsonian Institution, had crashed. Friends of Langley, who saw him as 'the father of flight', were unwilling to recognize the success of the Wright brothers, especially since they refused to publish their theory of aerodynamics. Eventually the historic aeroplane was lent to the Science Museum in London where it was exhibited until the end of the Second World War, after which it was returned to America and put on display at the Smithsonian Institution in Washington.

Frederick Lanchester (1868–1946)

The nineteenth century is regarded as the heroic age of British engineering, but there were some remarkable British engineers in the twentieth century. Hertha Ayrton and Charles Parsons have already been profiled; Frank Whittle and John Logie Baird are still to come. The mechanical engineer Frederick William Lanchester was a perfectionist, an engineer's engineer, whose unorthodox designs competed unsuccessfully with inferior orthodox designs. He was born at Lewisham, on the outskirts of London, on 23 October 1868. His father, an architect, moved his family to Brighton, on the south coast, where he built a house for his family. Frederick and his two younger brothers spent their boyhood and received their early schooling there. He won scholarships to Hartley College, Southampton, where his enthusiasm for engineering began to show itself, and then the National School of Science, South Kensington. Finding that this did not cover the practical side of engineering adequately he also attended evening classes at

the Finsbury Technical School. Although his ability was already recognized, he still lacked both academic qualifications and practical experience.

Lanchester first took a temporary job assisting a patent office draughtsman and before long took out his first patent, for a draughtsman's instrument. Then in 1889 he joined a small gas-engine works at Saltley, Birmingham, and within a year was appointed works manager. He soon reorganized the factory and designed a completely new gas-engine of greater size and power than that in production. He also improved the product by developing a pendulum governor and starter, which was a commercial success. Five years later, he installed his brother George as works manager and set up a separate firm of his own to exploit the commercial potential of his engine, which was now fuelled by petrol rather than gas. He first used it to power a motorboat, the first in Britain, but his aim was to produce a motor car.

When he went to see what was being done in France and Germany, Lanchester saw that all that was happening was to replace the horse in a carriage with a petrol engine, whereas he was convinced that it was necessary to design such a vehicle from scratch. He began by experimenting with a five-seater single cylinder 5 horse power model with a chain drive. This first took to the road one night in February 1896, at a time when such trials were against the law. Its successful 10-mile run so encouraged him that a second model, with a twin-cylinder 8 horse power engine at the rear and a new form of transmission, was built and tested in 1898; this won the Gold Medal of the Royal Automobile Club. A third model followed quickly, and led to the formation of the Lanchester Engine Company in 1899, which produced three or four hundred 10 horse power cars, cantilever sprung. Unfortunately, four years later the firm went bankrupt, and though a new company marketing cars bearing Lanchester's name was formed in 1905, he did not have much to do with it.

The design of aeroplanes, in those early days, was as much an art as a science. For some years, Lanchester had been working on the theory of flight, and his conclusions were published in his well-known two-volume book *Aerial Flight* of 1907 and 1908. In 1908, he met Wilbur Wright at Le Mans where the American was giving demonstration flights, and was impressed by his skill and daring, for flying was still quite dangerous. Six years later, he observed that practically all the distinctive features of the original Wright machine had disappeared; machines of that type were no longer made because of their unstable character. He was greatly in demand as a lecturer on aviation, and in 1909 was appointed to the Advisory

Committee on Aeronautics, chaired by the physicist Lord Rayleigh; whose responsibility was to initiate and direct research in that area.

The same year, he was elected president of the Institution of Automobile Engineers, founded in 1899. He was also appointed consulting engineer and technical adviser to the Daimler motor company, a position he held for the next 20 years, making many improvements to the design of their cars. However, most of his time and energy were spent on aerodynamics; the motor car took second place. After a brief flirtation with practical aeroplane design, he returned to the theoretical side of aerodynamics. On the outbreak of First World War, he wrote another influential book, called *Aircraft in Warfare, the Dawn of the Fourth Arm* (Lanchester, 2009), and this was followed by a long series of important technical reports for the various committees of which he was an active member. He was also a pioneer in what is now called operational research.

A number of commercially valuable inventions in automobile engineering had made Lanchester wealthy. During the war, he moved from Birmingham to London, where he took a seven-year lease on a large house in Bloomsbury. Music and sailing were his chief recreations. Lanchester was still a bachelor, but in 1918 when he was 51, he married Dorothea Cooper, the daughter of the vicar of a Cumbrian parish. After seven years in London they decided to move back to Birmingham, where they settled into a capacious house in the exclusive suburb of Moseley.

He was still primarily interested in aviation, especially the theory of flight, to which his old friends, the Göttingen mathematicians Runge and Prandtl, had contributed so much. Although he was still employed by Daimlers, there was increasing friction between the board of directors and their engineer. To solve this problem, a subsidiary of Daimlers was created to undertake development and research. In 1929, Lanchester acquired this company from Daimlers, so that he could undertake work for other manufactures. This used up most of his capital and before long the firm was in financial difficulties because it was hard to make a profit during a period of economic depression. Moreover, Lanchester began to have serious health problems. He appeared to make a good recovery from a major operation but physically he soon began to decline and Parkinson's disease was diagnosed. His intellect remained unimpaired and he continued to lecture and write, hoping to earn enough to make ends meet and avoid his house being repossessed. In the end, the Society of Motor Manufacturers came to the rescue, taking over his debts and providing him with a small pension. He lost his sight but consoled himself by writing poetry (published under the pseudonym of Paul Netherton-Herries). He died in Birmingham on 8 March 1946, leaving no children. He had been elected a fellow of the Royal Society in 1922; four years later the Royal Aeronautical Society awarded him a fellowship and their gold medal. The Institution of Civil Engineers awarded him another. He particularly valued the James Watt International Medal of the Institution of Mechanical Engineers, which he was awarded shortly before he died.

Guglielmo Marconi (1874–1937)

Guglielmo Marconi's central position in the history of radio development is due to a remarkable combination of inventive talent with a flair for picking out the practical features in the inventions of others and great business capacity. He was born on 25 April 1874 at the Palazzo Marescalchi in the ancient city of Bologna, the second son of an Italian father, Giuseppe Marconi, by his Scotch-Irish second wife, Annie Jameson (daughter of a well-known Dublin whiskey distiller). Marconi's formal education began at the Istituto Cavallero in Florence before enrolling at the Istituto Tecnico in Livorno, the great seaport and naval base on the west coast of Italy. In 1894, at the age of 20, he came across an account of Hertzian waves in a technical periodical. This led him to embark on a series of experiments at the Villa Grifone, his father's country house near Bologna, and within

12 months he had established wireless communication over a distance of more than a mile. He discovered, almost by accident, that an aerial and an earth were essential for good results.

Discouraged by the lack of interest in his work shown in his homeland, Marconi took his wireless apparatus to England, where his demonstrations created a sensation. Others, including the British physicist Oliver Lodge and the Russian physicist Alexander Stepanovitch Popov, had been working in the same field, but Marconi was ahead of them, more pragmatic, less theoretical. In June 1896, Marconi took out the first British patent for wireless telegraphy based on Hertz's discoveries. His apparatus, which used long waves, consisted of a tube-like receiver connected to an earth and an elevated aerial: its signals were at first transmitted over one hundred yards, a satisfactory demonstration being arranged from the roof of the General Post Office in London. Ship-to-shore communication was established in the following year, when Marconi formed a wireless telegraphy company in London for the exploitation of his patents in all countries except Italy. Once the Italian Government heard about Marconi's success in Britain, he was given a hero's welcome at home.

Many believed that wireless waves travelled in straight lines, like a beam of light, and would not bend round the earth's surface but Marconi was certain that the curvature of the earth would not be a problem for much greater distances. He believed that it was simply a matter of building a sufficiently powerful transmitter and in the last years of the nineteenth century he had shown this to be the case. Later it was found that long waves followed the earth's curvature, while short waves travelled in straight lines but were reflected by the ionosphere, the existence of which was established in the 1920s. In May 1897, communication was established in England between two naval vessels in motion while shortly afterwards communication was established between the Italian cruiser *San Martino*, when its hull was below the horizon, and the shore station ten miles away in La Spezia. Still better results were obtained during the British naval manoeuvres in the summer of 1899, when the best distance achieved was over 60 miles. Marconi realized the importance of not interfering with the ship-to-shore communication, which was proving so useful. Demonstrations were designed to afford valuable publicity for the company, which, for commercial reasons, kept its technical improvements secret. It was already clear that interference was a problem when several transmitters were in operation at the same time. The solution to this problem was to tune them to different frequencies.

In 1899, Marconi made his first visit to America to demonstrate his inventions and take out American patents. The Americans thought that interference would be a major problem but although Marconi knew how to overcome this, he had decided not to bring the latest tuneable apparatus because this was not yet protected by patents. Marconi often had to contend with industrial espionage. Altogether, the success of his visit was somewhat limited. However, on the voyage back to Europe he met and became engaged to an attractive young American woman. Two years later this was broken off; the 25 year old Marconi was in no hurry to get married.

Wireless transmissions across the English Channel were accomplished easily enough, once the French government had given its consent. However Marconi made history when the first transatlantic signals were sent in December 1901 from Poldhu on the coast of south-west Cornwall to St John's, the chief city of Newfoundland, where they were received from an aerial suspended from a kite. The demonstration came to an abrupt end when Marconi was handed a threatening letter from lawyers for the Anglo-American Telegraph Company, claiming that the company, which operated transatlantic cables, held a monopoly on all communications within the colony of Newfoundland. When this was known, the Canadian government at once offered him a site in Nova Scotia and a substantial grant towards the construction of a wireless station there, which could be linked to one at Cape Cod, which is in the United States. Efforts were then made to establish a commercial transatlantic service, but this took some years, due to technical difficulties; conditions for transmission were much better at some times than at others, owing, as we now know, to changes in the ionosphere.

Marconi was always careful to patent his inventions. In 1901, he patented selective tuning of radio waves, the next year, magnetic detection and, three years later, the horizontal directional aerial, of great importance to long-range wireless. In 1911, he took over the master patent of Oliver Lodge for a tuning device; and in the following year he introduced the time-spark system for generating continuous waves. In 1916, he began to experiment with very short waves in order to devise a beam system for military purposes; these researches were continued in peacetime and produced results that helped to transform long-distance wireless communication.

Marconi's company faced some major challenges. They wanted to establish a world-wide network of wireless communications, so they asked the British government for exclusive rights in perpetuity and other rights, arguing that the company had to expend a great deal of capital in developing the system. Influential allies lobbied on behalf of the company but the government was unwilling to grant it an exclusive monopoly, only an eight-year operating license. Other countries also raised objections.

Another headache was patent litigation, especially in the United States, and particularly with the American inventor Lee de Forest. Marconi transmissions were fit only for the Morse code; de Forest wanted to improve the quality so that music and voice could be transmitted. He invented a vacuum tube, which he called the audion, which amplified the signal and made this possible. Later, the Bell Telephone Company bought the patent and used it to amplify telephone messages. However, de Forest made other important inventions, including one for the first talking motion pictures. Through aggressive management and technical ability, his company was taking a good share of the market that the Marconi interests regarded as their own. He was cavalier about giving credit to his creative predecessors and made money for lawyers, if not for himself, either suing or being sued for patent infringement.

Famous, rich and single, Marconi enjoyed the social life of the Edwardian era, which revolved around house parties, and at one of these he met an attractive young woman, Beatrice O'Brien, daughter of an Irish peer with an estate in County Clare as well as a London house. When Marconi proposed marriage to her at first she turned him down. However, they continued to see each other and eventually they were married in the spring of 1905. Beatrice gave birth to one child, who died after a few weeks, then to a daughter Degna, and then in 1910 she gave birth to a son Giulio at the Villa Grifone, which had remained unoccupied since the death of Marconi's father in 1904. In London, they rented a large house, the Old Palace, in Richmond Park which made a good base for the active social life she loved. Before long, Marconi gave this up and moved to one called Eaglehurst in the New Forest. Although this provided a more suitable family home, the marriage did not prosper and they came close to a formal separation. In 1915, they left Eaglehurst and moved to Rome, where Beatrice was again able to enjoy the social life that was so important to her. Two years later he moved the family to the Villa Grifone, then in 1918 back to Rome.

Meanwhile, the Marconi Company was gaining fresh publicity, for example for the rôle of its equipment in helping to rescue survivors from the *Titanic* disaster of 1912. The Marconis would have been on the fateful maiden voyage themselves but for a last-minute change of plan. He took a later sailing because he needed to be in the United States at the time in connection with the takeover by American Marconi of the successor company to the de Forest Wireless Telegraph Company, which had been forced into liquidation as the result of a lawsuit for patent infringement. This was the occasion of the Marconi scandal, in which three of the Liberal government ministers were involved, although Marconi himself was not. Royal approval was shown when he was made an honorary Knight Grand Cross of the Royal Victorian Order.

In 1919, Marconi acquired a large and luxurious steam yacht, which he renamed the *Elettra*, on which he was able to continue his experimental work. He also used the yacht for entertaining important people and friends, especially attractive women. Bea did not like this and in 1923 she asked him for a divorce. When this was complete, she married again while Marconi, after several other liaisons, eventually married Cristina Bezzi-Scali, a quiet and serious Italian girl half his age. Her family was of the Roman nobility, her father highly placed in the Vatican, and approval was unlikely for her marriage to a middle-aged Waldensian Protestant, divorced with three children. It was essential that his previous marriage be annulled by the ecclesiastical court; somehow this was achieved. He never saw his first wife again, although he kept in touch with her children, who tried to achieve a reconciliation.

Marconi's marriage to an Italian bride in 1927 marked the start of a spiritual withdrawal back to the country of his birth. In 1930, Cristina gave birth to a daughter, who was christened Maria Elettra Elena Anna. For some years, Marconi had been suffering from heart attacks of increasing severity, although he continued to pursue his scientific and business interests. He died in Rome on 20 July 1937. In his will, he left everything to Cristina and Elettra, nothing to Beatrice and his children by her.

There is no Nobel prize for engineering. There is one for physics, more often awarded for theoretical rather than applied physics. In 1909, Marconi, as we know, shared it with Ferdinand Braun. In 1914, he was appointed senator, a title bestowed on mature and distinguished men of science and the arts. As a life member of the Italian Upper House he was often called upon to undertake diplomatic missions on behalf of the Italian government, including a place as an Italian delegate at the Paris peace conference. When Mussolini came to power, Marconi, who was a member of the Fascist party, was elected president of the new Royal Italian Academy, formed of people loyal to the régime. In 1929, he was elevated to the nobility as a marchese.

8 From Pal'chinskii to Zworykin

Peter Akimovich Pal'chinskii (1875–1929)

Russia has produced some great engineers, who had to contend with political as well as technical difficulties. The first to be profiled here, Peter Akimovich Pal'chinskii, became one of the victims of Stalin's paranoia. His father, Akim Fedorovich Pal'chinskii, a land surveyor and estate appraiser, married twice and had five children by his first wife, Aleksandra, seven by his second, Olga. Born on 3 October 1875, Peter was the oldest son by Aleksandra. As a child, he lived with his mother in Kazan, along with his brother Fedor and his three sisters Anna, Sophia and Elena, while the children of the second marriage lived with their parents in Saratov, further down the Volga.

Peter was an energetic youth and a bright student. After the age of eight, when his parents were divorced, he saw his father rarely. He confided primarily in his mother, a member of a socially prominent but impecunious family. She greatly influenced his early education. Under her tutelage, he became a good pianist, despite his lack of natural talent for the instrument. She also encouraged him to use the extensive family library, where, as well as works of literature, Peter read books on popular science and history. He also learned French and German; later he added English and Italian.

In the autumn of 1893, Pal'chinskii entered the Mining Institute of St Petersburg, one of the elite engineering institutions of tsarist Russia. He took special pride in the fact that he had been admitted on his own merits, without any help from influential friends or high officials. During his student years, Pal'chinskii was seriously short of money; the government grant he received was quite inadequate to meet his living expenses. To supplement this, he worked as a labourer on railways, in factories and mines during summer vacations. In these occupations, he developed a sympathy for manual workers and for the need to improve their pay and working conditions.

Like many young educated Russians around the turn of the century, Pal'chinskii was attracted to radical political doctrines that promised a better society than the authoritarian and poverty-stricken tsarist economy. As a result he attracted the attention of the secret police, who kept him

under almost constant surveillance from then on. Later he would be imprisoned five or six times, for political reasons. Compared with his siblings he was a monument of stability and prosperity; they often turned to him for help, psychological and financial. By his mid-20s he was supporting some of them.

On 23 November 1899, Pal'chinskii married Nina Aleksandrovna Bobrishcheva Pushkina, a member of a prominent liberal family of St Petersburg. They lived in the capital until he graduated from the Mining Institute the following year. As a student who had received a state stipend, however inadequate, he had to accept government assignments. In his case, he was directed to report on the reasons for the decline in coal production in the Ukrainian Don basin. The basin supplied over two-thirds of Russia's coal in 1900, so that inadequate supplies of coal threatened the continued growth of industry. Pal'chinskii was struck by the fact that the mine owners knew and cared very little about the mine workers living or working conditions, which were appalling. He sent back a report to St Petersburg, being careful to avoid political comment. Its impact was slow to take effect but when its implications were realized, he was sent to Siberia in what amounted to administrative exile, although permitted to continue to work as a consultant to the mining industry.

After the revolution of 1905, which he supported, Pal'chinskii was arrested and, although not sent for trial, exiled under police supervision to Irkutsk, in Siberia, under emergency powers granted to the police during revolutionary turmoil. While he was there, Pal'chinskii continued to work as an engineer and became an expert consultant on mining operations. He was valued by mine owners for his ability to improve productivity and reconcile differences between management and labour. However, he objected to police supervision and, in August 1907, he escaped Siberia and returned to the Ukraine, where he roamed from city to city to elude the authorities. Early the next year, he managed to slip across the border and begin a new life in Western Europe.

Meanwhile, his wife moved back and forth between Irkutsk and St Petersburg, unsuccessfully trying to persuade the authorities to drop the criminal charges against her husband. She was active in workers' education: she taught in special schools for workers, where she not only helped her pupils achieve literacy but taught them the political doctrines of reform and change. She was also concerned about the status of women and when, in 1909, she moved, with her mother, to join Pal'chinskii in Western Europe, she became active in the struggle for the emancipation of women and wrote articles about the women's movement for a feminist journal in St Petersburg. She led an increasingly independent life, once she discovered that her husband was unfaithful to her.

By this time, Pal'chinskii had become a successful industrial consultant, insisting on viewing engineering projects within their political, social and economic contexts. For example, when he was asked to improve the productivity and efficiency of major seaports he reported that it was not simply a matter of providing cranes, rail spurs, deep sea channels, wharves and warehouses; it was also a matter of workers' housing, schools, public transportation, medical care, recreational facilities, adequate pay and social insurance.

While Pal'chinskii seems to have adjusted well to life in Western Europe, he maintained his contacts in Russia. He wrote articles advising the tsarist government on how to improve the country's industry. The country, he maintained, needed hard-headed engineers who evaluated problems in all their aspects. The obstacles to Russia's industrial advancement, he believed, were not technological but political, social, legal and educational. For example, the legal system needed to be overhauled to bring order to land titles which, he wrote, were currently so disorganized that railways and mines were impossible to build because no-one knew who owned the land. He was particularly critical of engineering education, which he rightly believed was too academic. Pal'chinskii was convinced that Russia could sell coal and ores on the world market if it would only take the necessary political and economic steps.

In 1913, when his eight-year Siberian exile would have ended had he remained in Russia, he received a pardon from the tsarist government, and he and Nina returned to their native land. Three years later, he established an institute devoted to the rational use of the natural resources of Russia, which began publishing a journal about mining and industry. He served on the board of a mining company and established close connections with the business community. During the First World War, he was an advisor to the defence industry and served as deputy chairman of the government's War Industry Committee.

Pal'chinskii was a strong supporter of the provisional government that was established in Russia in February 1917 after the downfall of the monarchy. He held several official positions and supported the war effort against Germany. When the Bolsheviks took over the Winter Palace they imprisoned the top officials of the provisional government who had taken refuge there, including Pal'chinskii. Early the next year, he was released, then imprisoned again three months later, this time for nine months. At first Pal'chinskii, like the great majority of technical specialists in Russia immediately after the Revolution, had little sympathy with the Bolsheviks, whom they considered to be usurpers of power.

The Bolsheviks were committed to creating a command economy, to industrialization, and to science and technology. They seemed eager to benefit from the services of engineers and scientists. Pal'chinskii volunteered to help the new planning agencies that proliferated immediately after the Bolshevik victory and although dodging the police by staying away from Petrograd, as St Petersburg was now called, he consulted for a variety of Soviet offices and projects, including the building of the giant dam on the Dnieper river, the drafting of maps of population density and mineral deposits, and the construction of sea and river ports. He quickly became one of the best known engineers in Soviet Russia, serving as chairman of the Russian technical society and a member of the governing presidium of the all-Russian association of engineers. This did not prevent him being jailed for two months when he attempted to renew his youthful anarchist connections.

Pal'chinskii was eager to work with the Soviet authorities and the communist party in planning industry and increasing the strength of Russia but stoutly resisted the takeover by the party of any organization of which he was a member. His outspoken ways often got him into trouble. He was an independent, even stubborn, man who refused to give an evaluation of a project until he had studied all the relevant data. In 1926, he made a long tour of Soviet Central Asia, evaluating the potential of the oil and gas industry. With similar-minded colleagues, Pal'chinskii developed a programme for the industrialization of the Soviet Union, based on the exploitation of its enormous mineral riches. He thought that Soviet engineers, freed from capitalist employers, could have a greater influence on their nation than engineers anywhere else.

Although he favoured the general idea of central planning, Pal'chinskii considered that it should be combined with regional planning based on careful studies of local characteristics. He strongly criticized the belief of party leaders that the best facilities would always be the largest. He advocated foreign investment. Above all, he advocated humane engineering; concern for the needs of workers was not just an ethical principle but a requirement of efficient production. The Bolshevik leaders admired American methods of production: he did not.

When Stalin gained absolute control at the end of the 1920s, Pal'chinskii found that his ideas were considered dangerous. Stalin particularly mistrusted specialists, like himself, who were educated before the revolution and prepared to eliminate such people. In April 1928, a group of engineers was charged with sabotaging coal mines in the north Caucasus, some were acquitted, others sentenced to long terms of imprisonment or death. This was the start of a reign of terror among Soviet engineers. Early in 1929, Pal'chinskii was charged with treason and executed in secrecy without trial. His wife, Nina Aleksdrovna, was also arrested and disappeared into the camps.

Edith Clarke (1883–1959)

Until the end of the nineteenth century, and even beyond, there were all kinds of obstacles that made it difficult for a woman to become an engineer. Although women inventors were not uncommon, the doors of the profession were closed to women. The ability of Edith Clarke to survive in a male-dominated profession served as an inspiration to other women who aspired to a career in science and technology and helped to break down the barriers that had precluded women from pursuing such careers.

Born on 10 February 1883, she was one of nine children of John Ridgely Clarke, lawyer and farmer, and his wife Susan Dorsey Owings. Since her father died when she was seven and her mother five years later, an uncle

became her guardian while her sister looked after her and the other children. The family farm, where she spent her childhood, was in Maryland. She began her school education there, and then was sent to a boarding school up to the age of 16, where she received a conventional education; mathematics was her strongest subject. When she left school, she had no thought of a career but enjoyed the old-fashioned social life that still existed in the farming community of rural Maryland, relatively untouched by the American Civil War.

At the age of 18, she came into her inheritance and decided to use it to resume her education, against the advice of family and friends. After some preparatory lessons from a tutor in classics, she passed the entrance examination for Vassar college, where she majored in mathematics and astronomy. After graduating with honours she spent three years as a schoolteacher before deciding not to make a career of it. In 1911, having recovered from a serious illness, she enrolled at the school of civil engineering of the University of Wisconsin, where she lived in a sorority house and enjoyed the social life that it provided. At the end of her first year, she took a vacation job with the American Telegraph and Telephone Company (AT&T) in New York City.

At this time, the only openings for women in engineering in America were as numerical analysts, then known as computers, assisting male research staff with the tedious and time-consuming calculations that their work entailed. In Clarke's case, this was related to the transmission and distribution of electrical power. She became so interested in this that, instead of returning to Madison as she had intended, she took a permanent position at AT&T, where she trained and supervised a workforce of numerical analysts. To improve her qualifications, she took a course in radio at Hunter College and several at Columbia University, in the night school.

Clarke then left AT&T to study full-time at MIT, first as a senior undergraduate and then as a graduate student. In 1918, she received a Master's degree in electrical engineering, the first woman to receive such a degree from this prestigious school. However she found that this did not lead to the engineering position she was determined to get. She fell back on work similar to what she had been doing before, but this time in the turbine division of General Electric at Schenectady, New York State. After two years more of human computing she decided to satisfy a yearning for travel by becoming professor of physics at the Constantinople Women's College, in Istanbul. On the way there, she travelled through France, Switzerland and Italy, and on the way back, Austria, Germany, the Netherlands and England. During college vacations, she explored Egypt and the interior of Turkey, which she found fascinating. The urge to travel never left her; on leave in 1928 she went to the Arctic.

In her absence, General Electric reassessed her worth and on her return she was appointed a fully fledged engineer, at last, in the central station engineering department. Her work mainly consisted in solving the special problems encountered in large electrical power systems. Using her mathematical knowledge, she developed devices and charts by which many laborious calculations were rendered unnecessary. She wrote up her work in 18 technical papers published by the professional societies; these papers greatly simplified the analysis of transmission line and power systems by the use of applied mathematics. She also published a book on the subject. She invented a voltage regulator for long-distance power lines, which prevented an excessive drop in the terminal voltage, thereby making it possible for such lines to transmit more current by maximizing voltage.

When she retired from General Electric in 1945, she returned first of all to her native Maryland but, after two years, took up a position at the University of Texas teaching electrical engineering, which she held until 1956. The last years of her life were spent in Maryland, where she grew up; she died on 29 October 1959. Five years earlier, she had received the Achievement Award of the Society of Women Engineers. She was the first professionally employed female electrical engineer in the United States as well as one of the first women engineers of any kind. She was one of the earliest women elected to the American Institute of Electrical Engineering.

Andrei Tupolev (1888–1972)

Pal'chinskii was perhaps the most prominent of the Soviet engineers who were victims of the Stalinist purge. Most were thrown into labour camps with little chance of survival. The luckier ones, such as the subject of this profile, were placed in special research and development prisons and assigned tasks by the government. In the special science and technology facilities, often located within large cities, living conditions were comparatively good and they had relative autonomy, yet they had no doubt that they were in prison and were forbidden any contact with the outside world. The system was hopelessly inefficient; for example, the engineers who designed the White Sea Canal were not allowed to inspect the terrain through which it passed. Arbitrary acts of terror, such as the sudden removal of a prisoner from the workshop, might happen at any moment. The obsessive secrecy that operated in Soviet society slowed the flow of information, halting

projects at critical junctures. Stupid political operatives and secret police officers interfered with the engineers' work.

Despite this, remarkably good work was done, particularly in the most famous of these facilities, known as the Aviation Gulag, under the leadership of Andrei Tupolev. The aircraft designer was born on 10 November 1888 in Pustomazovo, a small town north of Moscow. His father, Nikolai Ivanovitch Tupolev, was a notary and subsistence farmer who had been an active radical during his student days at Moscow University. His mother, Anna Vasilyevna, née Lisitsina, had a facility for languages. An educated, middle-class family, they lived on a small farm near Tver (now Kalinin), the provincial capital, where their seven children were sent for their formal education. At the provincial gymnasium Andrei shone at mathematics and physics but not at other subjects. His handwriting was very bad but he read a great deal. He greatly valued his independence and personal freedom.

Against the advice of his father, his family and his school-friends, Tupolev decided to dedicate himself to the natural sciences. The first step was to pass the entrance examination for the Imperial Moscow Technical Institute, which offered the best technical training in Russia. There he came under the influence of Nikolai Zhukovsky, regarded as the father of Russian aviation, who was the first to provide a scientific explanation of the lifting force of an aerofoil, and to calculate its magnitude. He was inspired to pursue aeronautics, and gradually became Zhulovsky's assistant.

In 1911, there were student uprisings in Moscow, in which Tupolev was involved. He was arrested and imprisoned. While he was in prison, Tupolev's father died and he was released to return home and look after the family farm. In 1912, he was allowed to return to Moscow and resume his studies. Three years later, he was invited to be supervisor at a factory making flying boats, but Zhukovsky lured him back to join him in training the pilots of military aircraft. When the October Revolution was over, Zhukovsky was one of the older scientists who immediately pledged loyalty to the new Soviet government, and the rest of his team followed his example. Meanwhile, Tupolev gained his diploma with a thesis on the theory of the seaplane, which Zhukovsky praised highly. Encouraged by Zhukovsky, Tupolev then proposed the creation of a world-class Central Aerohydrodynamics Institute in Moscow, incorporating some of the existing facilities. The proposal was put before Lenin, who gave his approval, and the centre was established without delay. Tupolev was chosen to lead it, although his authoritative manner alarmed some of his colleagues. He was suffering from pulmonary tuberculosis, which obliged him to spend a year being treated in a good sanatorium. He returned much refreshed, determined to design his first real aeroplane.

In the purge of the engineers, Tupolev was not one of the first to be arrested and he was one of the lucky ones who were confined to the Aviation Gulag rather than sent to a labour camp. Living conditions in these prisons were tolerable but, even in this priority industry, engineers and scientists laboured under constraints that Western professionals rarely encountered. Tupolev was subject to interference by people who knew nothing about aircraft design. On one occasion he was summoned by Beria, the all-powerful head of the secret police. Beria wanted to know the specification of the dive-bomber he was working on. He then ordered Tupolev to increase the speed, the range and the load it could carry, and then dismissed him.

Under Tupolev's leadership more than 100 types of aircraft were designed, from light fighter planes to huge long-range passenger aircraft. To support the legitimacy of the régime, the Soviet government were promoting grandiose engineering projects of various kinds. As part of a campaign for technological display, one-of-a-kind aeroplanes were produced. The most remarkable of these was the huge *Maxim Gorky*, perhaps the largest plane in the world at the time, which was designed to break long-distance records. This appeared at international air shows until it crashed in 18 May 1935.

Tupolev was the first designer in the Soviet Union to use all-metal construction in both civil and military aircraft. In 1955, he built the first Russian jet passenger aeroplane, powered by engines imported from Britain. Yet his team of designers were forbidden the use of computers, which would have been invaluable to them. He was appointed a corresponding member and then, in 1953, a full member of the Soviet Academy of Sciences. He also received honours from other countries.

There were other remarkable aircraft designers in the Soviet Union, for example Ilyushin and Sikorsky. The latter, who designed flying boats and helicopters, was more original than Tupolev, whose team often copied designs that were developed in the West. American warplanes that crashed in Siberia were always carefully examined and their features copied. Tupolev spent the last years of his life working on the Tu-144, the Soviet Union's supersonic transport built to challenge the Anglo-French Concorde. When he died, on 23 December 1972, the Soviet press chronicled his many successes but did not mention that some of his aircraft were designed in the special prison workshop, whose existence was still a state secret.

John Logie Baird (1888–1946)

Scotland has given us Watt, Rennie and Bell; I now profile another Scot, one of the last of the private inventors. The youngest of the four children of the clever and fiercely independent Presbyterian minister John Baird, John Logie Baird was born at Helensburgh, then a small fashionable resort on the Firth of Clyde, on 13 August 1888. His capable mother, Jessie Morrison Inglis, came of a shipbuilding family in Glasgow; she gave birth to two sons and two daughters. In his boyhood, Baird, who was shy and short-sighted, had spent most of his leisure time either reading or constructing electrical gadgets. After a conventional school education, John took an engineering course at the Royal Technical College, now the University of Strathclyde, during which he gained some experience of monotonous hard factory work, and then went on to start a year at Glasgow University, which came to a premature end when the First World War broke out. Persistent ill health caused him to be rejected when he volunteered for military service and to resign from a post as a superintendent engineer of the Clyde Valley Electrical Power Company, which he held during the war years. Then he tried three small-scale commercial ventures in succession, with mixed success. The first involved selling under-socks in Glasgow, then preserves in Trinidad, to which he had gone for health reasons, and then assorted groceries in London. After a complete mental and physical breakdown in 1922, he retired to live in Hastings, a seaside resort in East Sussex. His mother died in 1924, providing him with a small legacy, which helped him embark on the work that was to make him famous.

Although handicapped by ill health and short of money Baird began trying to put his knowledge of electrical engineering to use. He was particularly interested in television, which after an experimental phase was awaiting someone to develop it into a commercial operating system. Efforts were under way in France and the United States but had not progressed very far since 1911, when the Scottish scientist Campbell-Swinton, in a public lecture given in London, had described the ways in which television might be achieved. There were two possible methods, Campbell-Swinton explained; one used mechanical scanning, the other electronic. He concluded that only a large American firm, such as Westinghouse Electric, would have the resources to make a viable system. Baird chose the mechanical method and after two years had contrived a primitive apparatus, capable of transmitting a flickering image over a distance of a few feet. Having conveyed his equipment to two attic rooms in London, he gave the first official demonstration of true television in January 1926. Fear of industrial espionage meant that he was extremely secretive about his process.

In December the same year, Baird showed his 'noctovisor', enabling images to be transmitted from a dark room by means of infrared rays. He also patented a form of fibre optics. In 1927, Baird demonstrated television by telephone line between London and Glasgow and the next year between London and New York, and also to a ship in the mid-Atlantic. He also pioneered television in natural colour, and stereoscopic and big screen television; and, in 1929, ultrashortwave transmission. He formed the Baird Television Development Company, which was allocated wavelengths allowing it to experiment with the first true television programmes. Synchronization of sound and vision was achieved a few months later, and half-hour programmes were transmitted regularly on five mornings a week. Because the receivers were very expensive, the programmes were generally shown in public places. The showing of the Derby horse race at Epsom in 1931 attracted a great deal of publicity. Foreign visitors were amazed at the uniformly high quality of the pictures, the regularly scheduled programmes, and the coverage of outside broadcasts, but the British Broadcasting Corporation, which enjoyed a national monopoly of radio transmissions in Britain, displayed a generally unhelpful attitude. Baird's company was also fiercely opposed by Marconi's firm, which was allied to powerful American interests, notably the Radio Corporation of American and its affiliates. As we shall see in the next profile, they had developed a competing system of television, known in Britain as the Marconi-EMI system.

Baird was now 40, and his affairs were beginning to prosper. He moved to a comfortable villa on Box Hill, in the country south of London. He had always received favourable press coverage in the United States and so he went there in 1931 to negotiate for a wavelength with the authorities and launch his system. He had already established links with France, Germany and other European countries. Although his visit was not a success in business terms, while in America he married Margaret, gifted daughter of the late Henry Albu, a manager at the De Beers diamond mine at Kimberley, who had come to London to study music and aspired to become a concert pianist. After their return to England he installed her first in his villa on top of Box Hill, then more suitably in another house in Primrose Hill, close to the centre of London. Then, without consulting her, in 1933 they moved again, to a very large Georgian house in Sydenham, convenient for his research activities at the nearby Crystal Palace, the huge cast-iron and glass edifice originally built in Hyde Park but subsequently moved to this part of south London. In 1935, a major fire at the Crystal Palace destroyed valuable equipment and records of his research.

Baird was waiting for the report of a departmental committee on which of the rival television systems were to be adopted as standard in the United Kingdom. Eventually, the committee came out in favour of the Marconi-EMI system. For two years, Baird's company operated side by side with its rival, which had the resources of the Marconi organization behind it. In September 1937, Baird's system, working on 240 lines, mechanically scanned, gave place finally to the electrically scanned 405 line system promoted by Marconi-EMI. Baird struggled on, but he had no income and had to live on his limited capital; his shares in the company were worthless. He had two children; a daughter Diana, born in 1932, and a son, Malcolm, born in 1935. When the Second World War began, the family moved to the small seaside resort of Bude, on the north coast of Cornwall, although most of the time Baird himself continued to live and work at Sydenham.

Baird continued to work on television, but using the technology based on the cathode ray tube rather than the scanning disc. He was also interested in the technology we know as radar, which was to play a vital rôle in the war and afterwards. Although the applications were very different, the two technologies were virtually identical. While Britain pioneered the development of radar, the United States, France, Germany and the Netherlands were not far behind. Although Baird offered his services to the War Office, he was not called upon in any significant way. As the war drew to a close, his health was failing, and his wife had moved the family to Bexhill, close to Hastings. In February 1946, he suffered a stroke and he died in the early hours of 14 June. He was survived by his wife and their two children: Malcolm, who became professor of chemical engineering at McMaster University, in Canada, and Diana, who was a schoolteacher before marriage.

It cannot be denied that Baird had defects of character. He could be charming in company but he made enemies, one of whom was Reith of the BBC. He did not have much business sense and did not realize the need to improve on his most promising inventions, rather than keep making new ones. His wife, in her memoir of him, recalled that his sense of humour could be cruel and wounding. What she found depressing was his lack of enthusiasm for anything except television. He had plenty of acquaintances, plenty of business contacts, but no real friends. One of his associates commented:

> My most vivid impression was his enormous toughness, underneath the quiet, dreamlike quality of his external personality. He would stop at nothing to achieve his end, which was always the furtherance of television. He had an unmatchable sense of humour and great courage but I shall remember his resilience till I die.

Another described him as having the vision of a prophet, the happy confidence of a child and the business sagacity of a sheepdog. Yet another described him as an eccentric visionary with a passion for gadgetry, a modest man of inflexible resolve, ready to try anything but constantly in financial trouble. Quiet, humorous, always approachable, he never made extravagant claims.

Vladimir Kosma Zworykin (1889–1982)

After the October Revolution, some of the leading Russian engineers emigrated to the United States. One of them was Vladimir Kosma Zworykin, Baird's rival in the development of television. He was born on 30 July 1889 in the ancient Russian town of Murom, the youngest of seven surviving children, five daughters and two sons, of Kosma and Elena Zworykin. His father was a local businessman, whose mansion at the centre of the prosperous town, only 240 kilometres from Moscow, was where the children grew up. After graduating with honours at the local Realschule, he enrolled at the St Petersburg Institute of Technology to study physics. In the summer vacations, he was sent to gain some practical experience of working in industry, at first on the railway, then at a steel plant and finally at a power station. During his third year, he was one of a group of students who were sent to Germany, Belgium, France and England to familiarize themselves with European industry.

The professor in charge of the physics laboratory, Boris Rozing, was working on electrical telescopy and 'seeing at a distance', an early form of television. Zworykin became his assistant. There were also French and German scientists working in this area. To remedy his lack of knowledge of theoretical physics, which was undergoing a revolution at that time, Rozing arranged for Zworykin to study first with the leading French physicist, Paul Langevin, in Paris and then at the Charlottenburg Institute of Technology in Berlin. Unfortunately, Germany declared war on Russia soon after Zworykin arrived there and it was only with difficulty that he succeeded in returning to St Petersburg. He was promptly drafted into the army, where he set up a radio station and spent most of his time coding and decoding radiograms.

In April 1816, Zworykin married a student of dentistry named Tatiana Vasilieff after a brief courtship, but the marriage was not a success. At this stage, he had risen to the rank of lieutenant and was based in St Petersburg, until he was sent to a small desert town named Turgai on the border with Chinese Turkestan, where the army was trying to suppress a rebellion. When this was over, he was moved to Moscow, where revolution was in the air, and then back to St Petersburg, just renamed Petrograd, which was in a state of turmoil. Eventually, his unit was sent to a place on the Dnieper river opposite Kiev, which was in German hands, and Tatiana joined him there.

After Zworykin was demobilized they went to Moscow but conditions were so chaotic that he decided to leave the country. There was no time to lose, as he was liable to be arrested as a former tsarist army officer. He travelled first to Archangel, where he arrived in August 1918 and managed to get an American visa. Onward travel was still difficult but in November, when the armistice had been signed, he was able to secure a passage to the United States via London. After a few months in New York, he decided to return to Russia, this time across the Pacific, but conditions in Russia were still so unsettled that he only stayed long enough to collect his belongings before going back to the United States, where he was to settle permanently. All this time, he had been trying to locate his wife. Eventually, he found that she was living with friends in Berlin and a few months later they were reunited. Before long she was expecting their first child.

Although Zworykin had difficulty in communicating verbally in English he could read and write the language well enough. The Westinghouse Electric Company had a policy of hiring recent Russian émigré scientists who had come to the United States after the October revolution. A team of their engineers was working on the development of a new amplifying tube to be used in receivers of broadcasts from the Westinghouse radio station in east Pittsburgh. Zworykin joined the team, but after a successful first year he was shocked to find that his pay was being reduced. He resigned and went to work for a small company in Kansas City at twice his previous salary. He liked it there, but soon the laboratory was shut down after he convinced the owners the process they were trying to develop would never work. Then Westinghouse came back with the offer of a job under much better conditions than before. This gave Zworykin the opportunity to work on what he most wanted, an all-electric cathode ray television. The difficulty was with the camera tube, not with the receiver.

When Zworykin had found a way of dealing with this, sometime in 1925, he demonstrated a working model of his design to the senior management of Westinghouse, who were not impressed. As a result, the television project was taken away from him and given to a believer in mechanical scanning. By 1928, Zworykin was working on three different projects for Westinghouse, a new, highly sensitive, photoelectric cell, a new method for recording sound on cinema film, and a new telefax transmitter and receiver. In 1928, he was sent to Europe, where he saw that in television the Europeans were ahead of the Americans. When he reported this to Westinghouse, he was given what he most wanted, carte blanche to produce a viable commercial system of television.

In 1933, Zworykin was back in Europe and this time he made an official visit to Moscow, the first time he had returned to his homeland since 1919. The next year he went again for six weeks to give a series of lectures on television. After several more visits, the Soviet trading corporation placed a large order with Westinghouse for radio and television equipment. In England, he heard that a public television station at Alexandra Palace would soon be functioning, although the receivers were costly. A comparable service was not started in the United States until 1939. A few months later, on the outbreak of war, the British transmission was shut down, and the American followed suit after the United States came into the war.

Zworykin had already taken out American citizenship at the first opportunity. In 1943, he had been elected to the National Academy of Sciences, shortly after moving to Princeton. In 1944, the Federal Bureau of Investigation became interested in his activities, and he was placed under surveillance. He was denied permission to leave the country, but in 1946 his file was closed and Zworykin was free to travel again. Doubts about his loyalties resurfaced in 1954, towards the end of the McCarthy red-baiting era, but an investigation failed to disclose any espionage activities or membership of the Communist party. In 1967, Zworykin was presented with the National Medal of Honour by President Nixon, 'for major contributions to the instruments of science, engineering and television, and for his stimulation of the application of engineering to medicine.'

Zworykin still spoke English with a heavy Russian accent. He had a wide range of non-technical interests and displayed the social graces typical of the well-to-do in tsarist Russia. Among his many friends were writers, artists, musicians and philosophers, especially Russian émigrés. In 1951, he divorced his first wife, after 20 years of separation, and married Katherine Polevitsky, a widow who was a neighbour at his summer home at Taunton Lakes, New Jersey. He died in Princeton on 29 July 1982, just one day short of his 93rd birthday.

9 From Gabor to Shannon

Dennis Gabor (1900–1979)

The future scientist, engineer, inventor, humanist and Nobel laureate Dennis Gabor, in Hungarian Gábor Dénes, was born in Budapest on 5 June 1900. The eldest in a family of three boys, he was followed by George, who died in 1935, and then André, born in 1903. Dennis knew his paternal grandfather, who had been born in 1832 of parents who had settled in Hungary at the end of the eighteenth century, having come from Russia and Spain. The family were tall, fair blue-eyed people, thought by the family to have been descendants of one of the Russian tribes, the Cerims or Kuzri, who adopted the Jewish faith centuries earlier. The boys' father Bertalan (or Bartholemew) came from the Hungarian town of Eger in 1867. He had been a gifted and ambitious child who hoped to go to university and qualify as an engineer, the profession followed by several other members of the family: unfortunately, his father's business failed and as a result he had to leave school early and take a clerical job at the age of 17. Nevertheless, he worked his way up and succeeded in becoming director of the largest industrial enterprise in Hungary. Their mother, Adrienne (née Kalman), was an actress who gave up the stage when she married. Her father was a highly skilled watchmaker and the son of an excellent tailor, but Dennis knew very little of his mother's forebears; he thought they were probably Sephardic Jews who settled in Hungary in the eighteenth century.

The three brothers grew up in a home of culture, where German was spoken as well as Hungarian, and the children were provided with French and English governesses in succession so that they became fluent in all four languages. In their early teens, the boys had intellectual stimulation not only from an excellent tutor but also from their father's circle of friends, including a doctor and a lawyer, who took a special interest in the talented young Gabors. The family took lunch and dinner together and, especially in later years, mealtimes were like the meetings of a discussion group.

Dennis was a voracious reader and was gifted with a prodigious memory. At about 12, his father offered him a prize if he could memorize

Schiller's long (430-line) epic poem *Das Lied von der Glocke*. Dennis won the prize and could still recite the whole poem in later years. He amused himself by translating Hungarian poems into perfect German, using the same metre as the original. At home, he benefited from a fine collection of histories of the visual arts, including coloured reproductions of famous paintings. He could name the painter of practically any item in a picture gallery. Dennis was also gifted musically and, having a marvellous memory and a good voice, could sing parts of most operas in their original language.

Family influence was so strong that, by comparison, his formal education had comparatively little influence on his development. The gymnasium Dennis attended included mathematics and the sciences in its curriculum. After some months, the form master complained that he had a neurotic child on his hands and suggested that Dennis be sent to a special institution catering for unmanageable children. His father realized that the problem was caused by the low standard in the class and persuaded the school to keep him. Dennis gained top marks throughout his school career. He was almost always ahead of the syllabus and did not treat his teachers with due respect.

The physics master was a remarkably conscientious man who constructed teaching aids not provided by the school. Dennis's knowledge of physics, even at that early age, was superior to that of his teacher, since Bertalan allowed the boy to buy almost any book he desired, including German textbooks of advanced mathematics and physics. As Gabor claimed, 'Before I went to university, I knew all about the mathematics I was to learn there and more electromagnetic theory than I ever learned at the Technical University in Berlin.' Although he never took a degree in mathematics, he was a more than competent mathematician, as is obvious from even a casual study of his life's work: all his inventive ideas were based on full mathematical analyses of the relevant details.

Dennis was a somewhat delicate child but at school he developed good physique and athletic ability; later in life, he also played a fine game of tennis. He joined the officer training corps and qualified in artillery and horsemanship. In 1917, after passing the Matura examination, which qualified him for university entrance, he was called up for military service. This took him to northern Italy, then occupied by the Hungarian forces. Characteristically, he took the opportunity to add Italian to his other languages. On demobilization, Gabor returned to Budapest.

The collapse of the Austrian monarchy after the war shattered existing economic and social patterns. Moreover, much of the wealth and population of Hungary was transferred to Romania. From being the second capital of the Austrian Empire, Budapest became just the principal town of a small country. In 1919, a communist régime took over, led by the extremist Béla Kun. Since a majority of its leaders were Jewish, a wave of anti-Semitism swept the country when the régime fell after four months. Under the Fascist regency of Admiral Horthy that followed, Jewish rights were curtailed, because Jews were associated with communism. This was racist anti-Semitism, not religious.

At this stage, the whole Gabor family adopted the Lutheran faith and Dennis remained in that faith for the rest of his life. In post-war Hungary, it was engineering rather than pure science that attracted people of talent and ambition, so that they might be in a better position to earn a living. However, Gabor found that he disagreed so much with the policies of the reactionary government of the time that, in his third year, he decided to leave the country and continue his studies in Berlin. When von Neumann, perhaps the most brilliant star in the constellation of Hungarian scientists who made the same decision, was asked for his opinion as to what contributed to the exodus of so many in the interwar period, he said that it was a coincidence of some cultural factors which he could not make precise:

an external pressure on the whole society of that part of central Europe, a subconscious feeling of extreme insecurity in individuals, and the necessity of producing the unusual or facing extinction.

In Berlin, Gabor decided to specialize in physics rather than engineering. He had been attending the famous Tuesday physics colloquium at the university, which became the high spot of his week. He was also attending the 'unforgettable' seminars on statistical mechanics conducted by Einstein. This confirmed his resolution to become a physicist, although he preferred to be described as an engineer. After taking a senior research degree in the subject, Gabor was recruited by the physics laboratory of the firm of Siemens and Halske but, when Hitler came into power, his contract was not renewed. He returned to Budapest, where he found that the political situation had deteriorated further. Several physicists he knew were already thinking of emigrating to England; when he was offered employment by the parent firm of Metropolitan-Vickers, he did not hesitate. The high voltage laboratory where he worked was located at Rugby, where he settled down happily, and within two years he had married the daughter of a local railway engineer. He remained there for the next 15 years but afterwards described it as a sterile period for research, compared with the excitement of Berlin. This was mainly because as an alien he was excluded from the secret research on radar being done at the laboratory, in preparation for the coming war.

However, in these years Gabor came near to inventing the electron microscope and later the laser. He took out patents on many of his inventions, some of which were commercially valuable. In 1942, he gave a lecture on electron optics in which he anticipated the work on holography that later earned him the Nobel Prize. In 1949, after 20 years working in industrial laboratories, Gabor moved into academia when he was appointed Reader in Electron Physics at Imperial College, London. There, during what he later described as some of the happiest years in his scientific career, he led a team of young research students. He was a workaholic but strictly at his own pace, often taking a nap at his desk. He gave special lectures but not regular courses and was spared administrative duties, although he acted as a consultant in industry.

In 1958, Imperial promoted Gabor to a personal Chair in Applied Electron Physics. In the first part of his inaugural lecture, he focused on scientific questions, including the feasibility of thermonuclear fusion, but in the second part, he began to speculate about the future of society. As he explained:

> The future cannot be predicted but futures can be invented. It was man's ability to invent that has made human society what it is. The first step of the inventor is to visualize by an act of imagination a thing or state which does not yet exist and which appears to him as in some way desirable. He can then start rationally arguing backwards and forwards until a way is found from one to the other. For the social inventor the engineering of human consent is the most essential and the most difficult step.

Gabor went on to develop his ideas in what proved to be his most popular book, *Inventing the Future*, which was translated into many languages. In it, he expressed critical views about communism, and so he was rather surprised to be elected an honorary member of the Hungarian Academy of Sciences. After Gabor reached retirement age in 1967, Imperial appointed him to a professorial fellowship, which enabled him to retain his old office and the secretarial provision he needed. In February 1969, he made what proved to be his last public appearance when he gave one of the traditional Friday evening discourses at the Royal Institution in London.

Increasingly in retirement he devoted his time to writing on social matters. He wrote, 'Now that my future is mostly behind me I am passionately interested in the future which I shall never see, but I hope my writings will contribute to a smooth passage into a very new epoch.' He applied his penetrating intellect to some of the problems of man's survival, problems created by the advance of technology; in a series of books and discourses. The Gabors usually spent their summers in Italy, where they had built a villa at Lavinio Lido, a pleasant holiday resort near Anzio, on the Mediterranean coast south of Rome. He was a founder member of the Club of Rome where people of diverse backgrounds met to study problems of natural resources, nutrition, environment climate, the third world, income distribution, etc. For some time, Gabor had been considered for a Nobel Prize, and in 1971 he was awarded this in physics for his pioneering work on holography. There followed a cornucopia of honorary degrees and similar marks of distinction. In 1974, Gabor suffered a severe cerebral haemorrhage, which left him unable to read or write; later he almost lost the power of speech, although his hearing and intellectual powers were unimpaired. At first, his physical health remained good but four years later, after enjoying summer in Italy, he became bedridden and died peacefully in a London nursing home on 9 February 1979, at the age of 78.

Sergei Pavlovich Korolev (1907–1966)

The engineer who was to be most closely identified with the Soviet space programme was born in the Ukrainian town of Zhitomir near Kiev on 12 January 1907. His father, Pavel Yakovlevich Korolev, taught literature at the local gymnasium. His mother, Maria Nokolaevna Moskalenko, from an old Cossack family, had married him as the result of parental pressure; it was not a happy marriage and it broke up when their son was three years old. Claiming, quite wrongly, that his father had died, Sergei's mother sent him to live with her parents at another town not far from Kiev. Although stubborn and argumentative, he enjoyed his schooldays. He was always neatly dressed, always excellently fed and always lonely.

When civil war broke out in the power struggle that followed the October revolution the family moved to the port of Odessa, the scene of much fighting during the struggle between the Red and White forces. As soon as Sergei was old enough, he attended a trade school where, although trained as a roofer, he also learned some mathematics and physics from able teachers. He was an accomplished gymnast, but his main interest was in aeroplanes. His ambition was to enrol in the Zhukovsky Academy in Moscow, but to start with he enrolled in 1924 at the Kiev Polytechnical Institute, where there was a sizeable aviation section. In 1927, he moved to Moscow and realized his ambition of enrolling at the Zhukovsky Academy, from which he graduated in 1930. The previous year he had his first experience of flight, piloting a glider, which he and another enthusiast had designed. Under the supervision of Andrei Tupolev, he now designed and constructed a light plane. In 1929, he received his certificate as a qualified aeromechanical engineer, and began work in the aircraft industry. He designed another light plane, which was fairly successful, but in 1931, after a period in hospital, he became interested in rocketry and two years later helped to found a rocket research group, which over the years trained many of the men responsible for the great early Russian achievements in space.

Some years previously, he had been attracted by a classmate named Lyalya Vincentini, of Italian ancestry, who was studying surgery. In 1931, they got married and in spring 1935 a daughter, Natasha, was born. Three years later, in the early hours of a June morning, Korolev was suddenly arrested and accused of subversion in a new field of technology. The evidence for this relied on the testimony of three of his colleagues, who had been arrested earlier and subjected to torture. Like most of the others imprisoned during the Stalin purges of the late 1930s, Korolev was given no trial but beaten and forced to confess to the trumped-up charges of his colleagues. In September, he was sentenced to imprisonment for ten years with disenfranchisement for five further years and his property was confiscated. His appeals were denied.

Korolev was moved around in the gulag system and by October 1939 was held in one of the most dreaded; a camp in the Kolyma area of far eastern Siberia, where several thousand prisoners died each month; and the annual death toll was 30 per cent. While Korolev survived, his health was permanently damaged. In 1939, his case was reviewed and the sentence reduced from ten years to eight. The following year, he was moved to the special prison in Moscow where Tupolev's team were designing military aircraft. Later he was moved to Omsk and then Kazan, but still as a prisoner and denied any contact with his family.

In late summer of 1945, Korolev, having completed his prison sentence, was commissioned a colonel in the Red Army and sent to Germany to join other Soviet experts who were gathering information about the German ballistic missiles. While the Americans benefited most from what they learned, and took the best of the German engineers to the United States, the Russians also recruited a substantial group of German technicians for their institute of rocket science. Eventually it comprised about 1,000 people, with equal numbers of Germans and Russians. They were handicapped by the lack of technical information, taken by the Americans.

By 1957, the Soviet Union had built intercontinental ballistic missiles to aim at targets in the United States, and the successful launch of the *Sputnik* in October showed that the Russians were well ahead of the Americans in space travel. This success was largely due to Korolev. He was responsible for the design of the satellite in which Yuri Gagarin made the first space flight in April 1961, and of the first rocket to orbit the moon and take photographs of its reverse side. By then, the Americans were catching up and pouring almost unlimited resources into space travel.

In December 1965, Korolev was diagnosed with an intestinal problem and in the course of routine surgery to remove a bleeding polyp a cancerous tumour was revealed. He never recovered from the operation and died in Moscow on 14 January 1966. He was buried in the Kremlin Wall, an honour reserved for Soviet citizens of exceptional distinction. In 1948, Korolev had divorced his first wife and married a second the next year. In 1953, he was elected corresponding member of the Soviet Academy of Sciences, and he was also awarded a Lenin prize. His success has been attributed to an unusual combination of personal and professional qualities. His official obituary stated that inexhaustible energy and talent as a research worker, splendid intuition in engineering, and great creative boldness in solving the most complicated scientific and technical problems were combined with great organizational abilities and high personal qualities. We would like to know much more but the true story of the life of this remarkable man remains to be told.

Sir Frank Whittle (1907–1996)

Frank Whittle has a permanent place in history as the original inventor of the turbojet engine, as described in his first patent, published in January 1930, when he was only 22 years old. His invention has revolutionized both civil and military air transport all over the world. He was born in Coventry on 1 June 1907, the first child of Moses and Sara Alice Whittle. He learned much from his father who, although virtually uneducated, having started work in a Lancashire cotton mill at the age of 11, had become a skilful mechanic and was a prolific inventor. He bought a small engineering company in 1916. Educated at council schools until the age of 11, Frank won a scholarship to the nearby Leamington College. There, his schoolwork was undistinguished because instead of doing his assigned homework he spent his spare time in the public library poring over texts on the theory of flight and practical flying. He decided on a career in the Royal Air Force (RAF). At first, his attempts to enrol as an apprentice were unsuccessful, because although he passed the written examination he failed the medical. However, on the third attempt he was passed and in 1924 at the age of 16, he began the three-year Apprentice Course at Cranwell.

In the last term of their course, cadets were required to submit a thesis on a subject connected with flying. Whittle called his 'Future developments

in aircraft design', declaring that, 'If very high speeds were to be combined with long range it would be necessary to fly at very great heights where the low air density would greatly reduce resistance in proportion to speed.' He was thinking in terms of 500 mph, at a time when the top speed was around 150 mph. He concluded that the conventional piston engine and propeller combination was unlikely to meet the power plant needs of the high-speed, high-altitude aircraft he had in mind. After passing out from Cranwell, he had a year's experience as a pilot officer before taking a course for flying instructors. He continued to develop his thesis. The idea of using gas turbines to drive a propeller had already been considered, but the idea using one to produce a propelling jet had not. Whittle discussed this possibility with one of the instructors at the flying school, who was aiming to become a patent agent. The Air Ministry thought his idea was impractical as there were no materials that would withstand the high stresses and high temperatures involved. Nevertheless, he took out a patent and, since this was not placed on the secret list, his idea was published to the whole world.

He tried to find a firm seriously interested in developing jet propulsion for aircraft but without success. Discouraged, he let his original patent lapse when the time came up for renewal. Meanwhile, he continued successfully with his service career. In the summer of 1934, he was posted to Cambridge to study for the mechanical sciences Tripos. While there, he was introduced to a firm of investment bankers who agreed to provide limited funding for him to start up a small company to undertake the development work. This company, called Power Jets, obtained the use of a disused factory from British Thomson Houston, a long-established company that manufactured industrial turbines.

In 1930, Whittle married Dorothy Mary Lee, by whom he had two sons. Whittle graduated with first class honours and was allowed to remain at Cambridge for a useful postgraduate year. When this was over, the Air Ministry agreed that he could work full-time on the jet engine; previously he had been limited to six hours a week. He was able to contribute an important improvement in turbine design, which greatly increased their efficiency. The story of the next few years is one of technical difficulties surmounted but financial difficulties increasing. Neither the government nor industry were particularly helpful. However, the impending war with Germany was beginning to create a new situation. The Air Ministry, which had not shown much interest in the project so far, began to do so.

It was known that the Germans were also developing jet propulsion for aircraft. In 1934, a young German physicist, Hans von Ohain, began working on it. In contrast to Whittle's experience, the project was taken out of his hands and handed over to established companies; the original inventor being sidelined. While the support that Whittle received was half-hearted, until the war came he was left in charge of the project. The German V-1, known as the flying bomb, used jet propulsion. After the war, most of the German engineers who had been working on the project left Germany; some stayed in Europe but von Ohain was recruited by the United States, like Wernher von Braun, and worked in the aircraft industry although not on jet propulsion.

When the Second World War broke out, a blanket of secrecy was imposed. Whittle was under enormous pressure and his health deteriorated under the strain. He was treated for the first of a series of nervous break-downs, apparently by electroconvulsive therapy. By the end of 1941, the engine had reached the stage where it could be installed in an aircraft and flight trials were completed without any problems. The engine was ready to be produced in quantity. All the British aero-engine manufacturers were interested and so were the Americans. In fact, several gas turbine projects were already under way in the United States in the late 1930s, apparently none of them derived from knowledge of Whittle's work. In Britain, Power Jets was taken over by the state, with Whittle retained as a consultant, and Rolls Royce took over the entire operation. In 1943, the United States entered the War and it was agreed that Britain would share the turbojet engine technology with them. Whittle was sent to America, where he was awarded the Legion of Merit in Washington and honoured in other ways. At home he had been elected a Fellow of the Royal Society and awarded a knighthood. After several spells in hospital, he was invalided out of the Air Force with the high rank of Air Commodore.

By then he was famous. He retired to Devon, where he received a stream of national, industrial and academic awards, including the Order of Merit. He divorced his first wife after they had been separated for 24 years. In 1976, he emigrated to the United States and married again. He held research appointments at the United States Naval Academy and lived not far away from Annapolis. Up to the time of his death on 9 August 1996, he was actively developing schemes for large supersonic passenger aircraft.

William Shockley (1910–1989)

The electrical engineer William Bradford Shockley was born on 13 February 1910 to an American couple living in London. His mother, May, grew up in the Wild West of the United States, and studied geology at Stanford University. In 1908, she had married William Hillman Shockley, an MIT-trained mining engineer, 22 years older than herself, who spoke eight languages and speculated in mines for a living. He came of an old New England family and chose MIT partly because several of his forebears had been instrumental in founding it. Before marriage, he travelled the world in a life full of enterprise and danger. After marriage, he and his wife lived a gay life in London, where they moved frequently from flat to flat. William, their only child, showed signs of autistic behaviour, making his parents' life miserable through violent temper tantrums, which they were unable to handle. In 1913, they returned to the United States, initially moving in with May's mother and stepfather in Palo Alto, but their peripatetic life continued, impelled both by financial considerations and an obsession with privacy, although they remained in Palo Alto.

William Junior was educated at home until he was eight and began formal education at a local public school. At ten he went to the local military

academy, where he did well, especially in physics, and in 1924 proceeded to Hollywood High School. His interest in science was stimulated by a neighbour who was a physics professor at Stanford University. The next year, his father died, just before his son gained admission to the southern branch of the University of California, now UCLA. By this time, they were living in Los Angeles. In 1928, Shockley transferred to the California Institute of Technology (CalTech), presided over by Robert Millikan. The Institute was enjoying a golden age, attracting first-rate students and faculty. The leading physicists of the world came to visit.

By then, the young man was proving a brilliant student, with some peculiarities. To maintain his physique he followed a strenuous exercise régime; he was also an accomplished amateur magician, and became famous for elaborate practical jokes. In 1932, he graduated from CalTech and moved for graduate work to MIT; the following year he married a fellow student at UCLA named Jean Alberta Bailey, just a year older than he was. They had a daughter, Alison, soon afterwards. Three months before graduation, he was hired by the Bell Telephone Laboratories, then located in Lower Manhattan.

The Shockleys lived nearby, close to Greenwich Village. He began to publish scientific articles and Bell patented some of his inventions.

In 1939, the Laboratory moved to Murray Hill, New Jersey, where it was the first industrial laboratory to be modelled on a college campus. It employed nearly 6,000 people, including 2,000 scientists and engineers. It was easily the largest and best industrial laboratory in the world, financed by Western Electric from the equipment it built based on the research carried out in the laboratories. In Shockley's case this research was in semiconductors, such as germanium and silicon, which neither conduct nor block electricity. Their conductivity, which varies with temperature and purity, had been studied at universities but without any commercial applications being considered, although researchers at MIT had built a radio using semiconductors in place of vacuum tubes.

The Shockleys had already moved to New Jersey, and their first son, William Alden, was born there in 1942. By then, however, Shockley was caught up in war work and was seldom at home. He needed to be in Washington, where he lived at the University Club. The development of radar had convinced the United States Navy that scientists like him could contribute usefully to the war effort. He was one of those who developed what became known as operational research. This was sufficiently important for Shockley to be awarded the National Medal of Merit in 1946. Shockley's only book, *Electrons and Holes in Semiconductors*, published in 1950, became a classic of twentieth century science texts. In 1951, he was elected to the National Academy of Sciences, and there were other professional honours.

By this time, his marriage was in serious difficulty. Shockley had become overpowering and at times violent towards his wife, as he had been towards his parents in childhood. In 1947, she had given birth to another son, named Richard, who turned into a difficult child. In 1953, Jean was diagnosed with uterine cancer, which was dealt with by major surgery. Shockley himself insisted that she also received deep radiation therapy. While this was happening, he announced that he was leaving her. He then had several affairs before finding a soul mate named Emily Lanning. She was a senior psychiatric nurse, unmarried at the time, who came from upstate New York, where her father ran an oil refinery.

At Bell Labs, Shockley left the transistor research team and worked on his own. His administrative responsibilities were increased but he was not promoted. He became anxious to leave and start up on his own, using his unrivalled knowledge of semiconductors. He began to sound out people who might know where he could get the capital he would need to start up his own company, to be called the Shockley Semiconductor Laboratory, later renamed the Shockley Transistor Corporation. He secured a suitable backer, and founded a laboratory of his own, to which he recruited some exceptionally able researchers.

He divorced Jean and married Emily in 1955; she remained loyal to him throughout the vicissitudes of the rest of his life. They agreed that if either of them wanted to end their marriage they only had to say so. He was appointed to a lectureship in the flourishing department of electrical engineering at Stanford, running a graduate seminar on solid-state physics. Although an excellent teacher in other ways he was a hopeless lecturer. He ignored the difference between written language and speech, lectured in exactly the style he would use in a scientific paper no matter what the audience, and read in a monotone, without any inflection.

He bought a house in Los Altos, just south of Palo Alto, where his mother May had become moderately wealthy from investments left by her late husband and others she made herself. The adjacent Santa Clara valley was the ideal place for Stanford to establish an industrial park, and Shockley's firm was one of the first in the area. In 1956, he was awarded the Nobel Prize in Physics for the invention of transistors, along with John Bardeen and Walter H. Brattain, two of his former colleagues in Bell Labs. In fact his rôle in the invention was debatable. Undoubtedly, Shockley was the leader of the group that designed and constructed the first working model. He had the general idea but otherwise he was not directly involved in the construction of this model. However, as soon as the concept was shown to be practicable he designed the ancestor of the transistor as we know it today.

Shockley was recognized as a scientific genius but a poor businessman. He directed his firm to concentrate on the semiconductor silicon, not germanium. This was not a mistake in itself, but he went on to insist that it was used for telephone switches, not transistors. Unfortunately, the only potential customers for semiconductor switches were the large telephone companies, who required total reliability if they were to change from the mechanical switching that served them well. Shockley's firm tried to achieve this but until they could do so there were no customers for its product.

Shockley considered himself an expert at managing creative institutions and creative people. He wrote several articles on the subject, in which he expounded his belief that scientific advancement was due to just a few creative geniuses, like himself, whose ideas fed a lower layer, an intelligent team of researchers, who would develop them but would not be expected to have ideas of their own. In truth, Shockley had no idea how to manage. He competed with his own staff; screamed insults at them. He lost his temper with his financial backer, an honourable man, who was naturally concerned that the company had no customers and no income. His staff mutinied, insisting that either Shockley had to go or they would leave. Efforts were made to find a compromise but eventually the leading rebels split off and formed their own company, the Intel Corporation, which became the largest computer chip manufacturer in the world.

After Shockley left industry, Stanford appointed him the first Alexander M. Poniatoff Professor of Engineering and Applied Sciences, the chair he held from 1962 to 1976. In 1961, the family were involved in a car crash, caused by a drunken driver, which left Shockley, his wife and younger son seriously injured. They never recovered completely. Apart from using an exercise machine, swimming became his main form of exercise; at the university pool he would race other swimmers whether they liked it or not. He had to give up his favourite sport of rock climbing, because of his physical injuries, but took up sailing instead. The family moved into a select residential area reserved for permanent Stanford faculty and top administrators. To supplement his income, Shockley obtained part-time work at Bell Labs, teaching their interns.

The controversial policy of the hereditary improvement of the human race by selective breeding, rather than leaving it to natural selection, is known as eugenics. Positive eugenics, such as the encouragement of early marriage between talented men and women, should be distinguished from negative eugenics, discouragement in the case of those considered unfit. In Britain, public opinion was generally against the idea of negative eugenics but this was not so in the United States, where, by 1914, legislation had been passed in 30 of the states forbidding the marriage of the 'mentally deficient'. In Nazi Germany, the sterilization of those suffering from eight allegedly hereditary disorders was made compulsory, and it is estimated that about 400,000 sterilizations were carried out. Later, sterilization was replaced by euthanasia as being more cost-effective. Today, although the idea of negative eugenics is discredited, the idea of positive eugenics is by no means dead.

Shockley now began to advocate negative views on genetics at conferences to which he was invited as a famous Nobel Laureate. He maintained that three things threatened humanity: nuclear war, famine and, finally, 'genetic deterioration of the human race through lack of elimination of the least fit as the basics of continuing evolution'. He received considerable, if faint-hearted, support in the scientific community but he let his opponents assume the moral and scientific high ground. He donated sperm to a sperm bank.

In the early 1970s, a Stanford psychiatrist told a reporter that he thought Shockley was suffering from the classic symptoms of paranoia. Shockley taped ordinary and telephone conversations and saved every piece of mail, however trivial. He depended on his wife, Emily, who acted as his loyal secretary and had decided to devote her life to him. They had very few friends left; when he invited people to dinner, he would leave them at the table and return to his office down the hall. Emily never recognized his insensitivity, and seemed incapable of interpreting his behaviour as ill-mannered or unbalanced. She described him as a very warm, sensitive and perceptive person, but no-one else did. Most people thought that his paranoia and insensitivity made spending time with him more unpleasant than he was worth.

In 1977, his mother May died at a nursing home, at the age of 98, and his first wife Jean died in the same year from a recurrence of cancer. In 1982, Shockley ran in the Republican primary for a seat in the United States Senate, largely for the publicity it provided for his racial theories. He believed in polygraph lie detector tests and proposed that presidential candidates should take them to see whether they were telling the truth in their campaigns. Five years later, he was diagnosed with prostate cancer that had spread and on 12 August 1989 he died. His behaviour as a child, and after middle age, suggests a personality disorder on the autistic spectrum, which he might have inherited from his mother.

Wernher von Braun (1912–1977)

The controversial rocket scientist Wernher von Braun was born on 23 March 1912 in the town of Wirsitz, in the Prussian province of Posen. He was the second and most gifted of three sons born to Baron Magnus Alexander Maximilian von Braun and his wife, Baroness Emmy von Quistorp von Braun. The family owned estates in both East Prussia and Silesia. All the boys automatically became barons at birth, under Prussian law; they had the status of Junkers. Their wealthy father had farming and banking interests and was an official first in the provincial and then in the national ministries of agriculture: by 1924, he was Reichminister of agriculture during the Weimar Republic. Their mother, who was of aristocratic Swedish-German

lineage, was said to have possessed a brilliant mind and was able to get on well with other people. Brought up in England, she fostered a cultured home-life, where music, art and literature were important. Wernher was musical; he studied piano under Paul Hindemith and composed a little. Like his mother he was gregarious by nature.

The family had moved to Berlin in 1920. Wernher was sent to a progressive boarding school where a strong emphasis on academic subjects was combined with manual work. Although he liked schoolwork, especially mathematics and physics, he was not outstanding academically. He was already enthusiastic about rocketry and, when he left school, became a protégé of Hermann Oberth, an ethnic German born in Romania who had recently come to Berlin. When Oberth returned to Romania, von Braun persuaded other enthusiasts to join him in forming a company to continue rocket research in Berlin. Rocket science was outside the disarmament provisions of the Treaty of Versailles. The German army showed enough interest in what they were doing to give them some financial support. Meanwhile, von Braun was studying mechanical and aircraft engineering at the Charlottenburg Institute of Technology, after which he transferred to the University of Berlin, where he obtained his doctorate in physics.

In 1934, the year after Hitler came to power, success with their experimental rockets brought his company more substantial funding, from the Luftwaffe as well as the army. At the age of 21, von Braun was called up for military service in the Luftwaffe, during which time he obtained a regular pilot's licence. Next, he resumed rocket development at Peenemünde, on a sandy wooded island about 180 miles north of Berlin. Led by Hitler, the country was gearing up for war. The army wanted rockets that would carry warheads, was willing to pay for them and provided the necessary manpower. Of course everything had to be done in complete secrecy. The team constructed a 46-foot tall 14-ton liquid propellant rocket weapon designed to carry a 1-ton warhead over a range of about 210 miles. The test launch of this first ballistic missile came in 1942. The first two tests were failures but the third was not. Hitler ordered mass production, but assigned it low priority.

Long before, von Braun had been under pressure to join the Nazi party, and had done so in 1937, accepting a commission in the SS, rising to the rank of major. In 1943, Heinrich Himmler seized control of the project for the SS and rushed it into mass production before essential development work had been completed. The team were ordered to build 30,000 rockets, at a rate of nearly 1,000 a month, using slave labour. When they were ready, London was chosen as the primary target, then Paris and Antwerp. Over 3,000 rockets were fired successfully, resulting in over 5,000 deaths.

At the end of the war in Europe, von Braun's team decided to surrender to the Americans, who were already hunting for them. At first, the officer they surrendered to found it incredible that someone who seemed so young could have led such a programme. Nevertheless, it was confirmed that in rocket science the Germans were far ahead of the Americans and preparations were begun for moving operations to the United States, using material from Germany and members of von Braun's team. One hundred of the key engineers were chosen to continue their work in America; many others were keen to go. The Soviet Union was also collecting German rocket scientists and putting them to work near Moscow, very much against their will. Britain did not seek to retain any of von Braun's team but assembled and successfully launched three rockets over the North Sea.

The Germans, led by von Braun, ended up at Fort Bliss, Texas, and were employed on five-year contracts. At first their presence in the United States was kept secret. In 1947, von Braun got married, to a Quisborn cousin he had known for many years. He was allowed to return to Germany for the wedding, though under escort because of the risk that he might be kidnapped by the Russians. When he returned to America, he brought with him not only his bride Maria but his ageing parents, who had lost everything at the end of the war. Although he was settling down well to life in the United States, he was frustrated by the lack of science research facilities and other resources for his team. However, all this changed in 1949, when the Soviet Union tested nuclear weapons and intelligence reports revealed that its ballistic missile development programme was well advanced. The American army was ordered to produce a 200-mile, nuclear-capable missile on a high priority basis, followed by longer range weapons. The German team were moved to the Redstone Arsenal, at Huntsville, in verdant northern Alabama, and provided with everything they needed.

The von Braun's first child, Iris Careen, was born in 1948 when they were still at Fort Bliss. Their second, Margrit Cécile, was born in 1952 at Huntsville. The children automatically became American citizens, having been born on American soil. The parents were naturalized in 1955, after renouncing their German citizenship. A son, Peter Constantine, was born in 1963. On trips, von Braun rarely carried any money, leaving any payments to be made by an aide. He often fitted the stereotype of the absent-minded professor, was hopeless at dealing with everyday gadgets and machines, and ignored highway regulations but loved to fly aircraft. He was a fascinating conversationalist, extremely handsome and socially charming.

In 1970, the popular von Braun was given an ecstatic farewell by the citizens of Huntsville, all except the Jews among them, and moved to NASA headquarters in Washington DC. In the capital, he enjoyed an active social life; but there were difficulties at NASA and von Braun was never made head of the organization. In 1972, he left to work for Fairchild Industries; not long afterwards he was told he had colonic cancer. His condition rapidly worsened, despite major surgery, and by 1977 he was terminally ill. President Ford awarded him the National Medal of Science. He died on 16 June 1977.

Claude Shannon (1916–2001)

For the last profile I have chosen the American Claude Shannon, who might be described as a mathematical engineer. According to his obituary in *The New York Times*:

> He single-handedly laid down the general rules of modern information theory, creating the mathematical foundations for a technical revolution. Without his clarity of thought and sustained ability to work his way through intractable problems, such advances as email and the World Wide Web would not have been possible.

Although he is regarded as the founding father of the information age, he was not himself much interested in the applications of his theoretical work.

Claude Shannon was born on 30 April 1916 and grew up in Gaylord, a town of some 3,000 inhabitants in the northern part of Michigan. Like his elder sister, Catherine (1910–2009), he was born in hospital at the nearby town of Petoskey. His paternal grandfather was a farmer and inventor, while his father (1862–1934), of the same name, was clever mathematically and worked as judge of probate in Gaylord. The descendant of New Jersey settlers, he married Mabel Wolf (1880–1945), the daughter of German immigrants, who was a language teacher and for a number of years principal of the high school in Gaylord. Claude Shannon Junior was educated at the same school. In boyhood, he already showed mechanical aptitude: he tinkered with radio sets given to him by his father and mathematical puzzles provided by his sister. His hero was Edison, with whom he had a common ancestor.

After leaving school in 1932, Shannon entered the University of Michigan at Ann Arbor, following his sister, who had just obtained a master's degree there in mathematics and who went on to teach mathematics at college. In 1936, he obtained a first degree in electrical engineering and another in mathematics, the two subjects he was to pursue throughout his life. He joined the department of electrical engineering at the Massachusetts Institute of Technology (MIT) as a research assistant, where he worked with Vannevar Bush on one of the early analogue computers, the differential analyzer, which used a precisely honed system of shafts, gears, wheels and discs to solve equations in calculus. After two years he transferred to the department of mathematics, where he wrote a brilliant master's thesis in which he showed how Boolean algebra could be used in the design of relay and switching circuits, not knowing that Victor Shestakov, at Moscow State University, had proposed a similar theory much earlier, because this was not published until 1941. In 1940, after a summer at the Cold Spring Harbour oceanographic research station on Long Island, Shannon completed his Ph.D. dissertation, entitled *An Algebra for Theoretical Genetics*. This overlapped with some of J. B. S. Haldane's early work on population genetics, of which he seemed to be unaware, and although an excellent thesis it was not published until 1993, by which time most of his results had been obtained independently by others.

After spending summer vacations working at the Bell Telephone Laboratories in Manhattan, Shannon was recruited as a research mathematician, a position he held for 15 fruitful years. Many outstanding engineers were his colleagues, among them John Pierce, known for satellite communication, Harry Nyquist, with numerous contributions to signal theory, and

Hendrik Bode, of feedback fame. There were also physicists of the calibre of John Bardeen, Walter Brattain and William Shockley.

In 1939, Shannon published his most important paper, 'A mathematical theory of communication'. This seminal work transformed the understanding of the process of electronic communication by providing it with a mathematics, a general set of theorems rather misleadingly called information theory. The information content of a message, as Shannon defined it, has nothing to do with its inherent meaning, but simply with the number of binary digits that it takes to transmit it. Thus, information, hitherto thought of as a relatively vague and abstract idea, is analogous to physical energy and can be treated like a measurable physical quantity. Shannon's definition was both self-consistent and unique in relation to intuitive axioms. To quantify the deficit in the information content in a message, Shannon characterized it by a number, the entropy, adopting a term from thermodynamics. Building on this theoretical foundation, Shannon was able to show that any given communication channel has a maximum capacity for transmitting information. The maximum, which can be approached but never attained, has become known as the Shannon limit. So wide were its repercussions that the theory was described as one of man's proudest and rarest creations, a general scientific theory that could profoundly and rapidly alter man's view of the world. Few other works of the twentieth century have had a greater impact; Shannon altered most profoundly all aspects of communication theory and practice.

In 1940, Shannon was awarded a National Research Fellowship to work at the Institute for Advanced Study in Princeton, under the German émigré, mathematician Hermann Weyl. He was probably more influenced by the Hungarian mathematical polymath John von Neumann, whose many interests included computing, automata and game theory. During the Second World War, Shannon's official responsibility was in developing cryptographic systems. In 1943, he met for the first time his British opposite number Alan Turing, who was briefly visiting the Bell Laboratories. Turing's seminal 1936 paper, introducing what is now known as the universal Turing machine, contained ideas that were complementary to his own. However, under wartime conditions of secrecy, what might otherwise have developed into a fruitful collaboration could not progress very far. Shannon's work on encryption led to the system used by Winston Churchill and Franklin Roosevelt for transoceanic conferences, and inspired his pioneering work on the mathematical theory of cryptography.

On 29 March 1949, Shannon married Mary Elizabeth Moore, from Staten Island, who after graduating in mathematics at Rutgers University was working as a technical assistant in the microwave research department of the Bell Laboratories. They raised three very active children, Robert James (1952–98), who became a computer scientist, Andrew Moore (1954–), who became a pianist and composer, and Margarita Catherine (1959–), who became a college teacher. In 1958, Shannon was appointed Donner Professor of Science at MIT: in this capacity he supervised three doctoral students. He remained on the faculty for the next 20 years, although he maintained his connection with the Bell Laboratories until 1972. He mainly worked at home, a large house overlooking Mystic Lake in Winchester, Massachusetts, where students were always welcome and were frequent visitors. In the basement, a fully equipped machine shop, he spent much of his time making electrical and mechanical machines, for example, a cable railway that ran to the lakeshore. The house also contained numerous musical instruments, including five pianos; he played the clarinet himself and enjoyed Dixieland jazz. He formally retired in 1978, but continued happily working as before. Eventually, however, he succumbed to Alzheimer's disease, and died on 24 February 2001, at the age of 84, survived by his wife, their three children and his sister.

Although highly regarded professionally, at home and abroad, to the general public he was best known for the ingenious machines and gadgets he invented and built. Some of these were of real scientific value, for example, his mechanical 'mouse' Theseus, which learnt the path through a maze, was an early contribution to the field of artificial intelligence. There was also a 'mind-reading' machine, which could anticipate, more than half the time, whether a challenger would choose heads or tails. Other ingenious gadgets he invented just for the fun of it. For example, there was THROBAC, a calculator that operated in Roman numerals. Juggling was another passion of his. On one famous occasion, he rode a unicycle through the corridors of MIT while juggling several balls in the air, a stunt normally performed by circus clowns. A collection of these machines can be seen in the MIT museum.

According to an anonymous obituary in *The Times* newspaper,

> He was a playful genius who invented the bit, separated the medium from the message, and laid the foundations for all digital communications. He single-handedly laid down the general rules of modern information theory, creating the mathematical foundations for a technical revolution. Without his clarity of thought and sustained ability to work his way through intractable problems, such advances as email and the World Wide Web would not have been possible. Something of a loner throughout his working life, he was individually

> responsible for two of the great breakthroughs in understanding, which heralded the convergence of computing and communications. To colleagues in the corridors at MIT, who used to warn each other about the unsteady advance of Shannon on his unicycle, it may have seemed improbable that he could remain serious for long enough to do any important work, yet the unicycle was characteristic of his quirky thought processes, and became the topsy-turvy symbol of unorthodox progress towards unexpected theoretical insights. The ability to make astonishing leaps beyond the intellects of his colleagues (all the more remarkable at MIT, the forcing-house of technological theory) had to be acknowledged as genius, and Claude Shannon was recognized as a giant throughout the industry.

In an interview he gave in 1984, Shannon explained that he was attracted more to problems than applications. He just liked to solve a problem, an interesting problem, and didn't care whether someone else had been working on it or not. His rather offhand approach to problems was not always popular with his rather earnest colleagues, who felt he lacked the vigour of the true researcher. Like Babbage, he was known as the irascible genius; because he was not always patient with those who did not readily follow his line of thought. Yet normally he was a shy man, gentle and quiet, with a whimsical sense of humour. Shannon was little interested in the applications of his scientific contributions, which were often capable of successful commercial exploitation. Shannon became wealthy not from this but from carefully chosen investments in technological companies.

As well as his publications in scientific and technical journals, Shannon wrote articles for the 14th edition of the *Encyclopaedia Britannica* on information theory and on cybernetics. Norbert Wiener, regarded as the father of cybernetics, was his colleague at MIT; Shannon had great respect for him. He also edited and contributed to several books, but never wrote one himself. His publications essentially come to an end in 1967, when he was just over 50, although there are a few unpublished typescripts from later years and he was by no means inactive.

In 1983, Shannon looked back on the development of information technologies in his lifetime, writing:

> The growth of both communication and computing devices had been explosive in the last century. It was about a hundred years ago that the telephone and phonograph were invented, and these were followed by radio, motion pictures and television. We now have vacuum tubes,

> transistors, integrated circuits, satellite communication and microwave cable. We have even talked to astronauts on the moon. Our lifestyle has been totally changed by advances in communication. On the computing side, we started in the twentieth century with slide rules and adding machines. These were followed in quantum jumps by Bush analogue computers, Stibitz and Aiken relay computers, Eckert and Mauchly vacuum tube machines, transistor computers and finally incredibly compact integrated circuit and chip computers. At each step, the computers became faster, cheaper and more powerful. These hardware revolutions were marched by equally impressive developments in programming.

Shannon was Vanuxem Lecturer at Princeton in 1958, Gibbs Lecturer of the American Mathematical Society in 1965, Semi-centennial Lecturer at Rice University, and gave other prestigious lectures. After retiring from MIT, he spent Trinity term 1978 as a visiting fellow at All Souls College, Oxford, during which he gave one of the Chichele Lectures and the University conferred on him an honorary degree, one of over a dozen he received from various universities at home and abroad. He received numerous other professional and scientific honours, including the National Medal of Science, in 1966, the Kyoto Prize, in 1985, and other awards by which the engineering profession makes up for the absence of Nobel Prizes. He was elected to the American Academy of Arts and Sciences, the National Academy of Sciences, the National Academy of Engineering, the American Philosophical Society and the Royal Society of London, to which he was elected in 1991. Six copies of a portrait bust, by Eugene Daub, have been placed at various locations in the United States, including Gaylord, Michigan, where he grew up, the University of Michigan, where he was an undergraduate, MIT, where he was a graduate student and professor, the San Diego campus of the University of California, where he was a fellow, and the Bell Laboratories, where he spent his most fruitful years and where, after the forced break-up of the Bell Corporation, the remaining part of its laboratories where Shannon worked was named after him. Another is at the remaining part of the Laboratories, now owned by Alcatel-Lucent. Both the bust and a painting, also at the Shannon Laboratories, were based on a photograph taken in his lifetime.

Epilogue

As we have seen, the kinds of people who were attracted to engineering as a career were quite various. Some, such as Vauban, Brindley, Telford, the elder Stephenson, Ayrton and Woods, and perhaps also Edison, grew up in the shadow of poverty. Several, such as Riquet, Trevithick, Marc Brunel, Diesel and Lanchester, fell seriously into debt in the course of their careers. A few, like Cayley, Parsons and von Braun were born into wealthy families. They differed greatly in their social background and their degree of education. Engineers, however distinguished, were seldom accepted into the scientific academies. Instead they formed societies or other bodies to regulate their branch of the profession. Four engineers, namely Braun, Marconi, Gabor and Shockley won Nobel Prizes. Telford, the younger Brunel, Bazalgette and Parsons were knighted; Vauban, Thomson and Marconi were ennobled. Cayley and von Braun inherited a title.

The job of the practising engineer varies enormously according to the speciality. The work of a civil engineer, for example, might involve planning, costing and organizing the construction of somcthing, dcaling with legal problems and managing a work force. A mechanical engineer, might expect to design, manufacture, install and maintain machinery. There are also different ranks, different levels of skill, knowledge and responsibility, and different official or unofficial levels of status. However, any engineer requires a good business sense if he is to succeed. He also needs to develop managerial skills, to deal with contractors and employees, both skilled and unskilled. Several of my subjects were conspicuously poor managers.

Not all inventors are engineers, of course, and not all engineers are inventors, but many are. Producing and marketing a successful invention and fending off competitors is difficult. In the eighteenth century, an inventor would try and do everything himself; in the twentieth century most of these functions would be delegated, either to an outside contractor or to another member of the firm the inventor was working for. Raising money for projects was important for some; people had to be persuaded to invest and investors had to be kept on board. Few had enough capital to be able to finance themselves. They needed to take into partnership others who had money but who also may have had different objectives. Ideas are relatively cheap; it is the cost of developing them into commercial propositions that is expensive.

The inventor has to choose between trying to keep his invention secret, to avoid industrial espionage, or protecting it by means of a patent. Intellectual property rights have a long and complicated history, and the situation varied from country to country and from time to time. In Britain, the Crown's abuse of letters patent to reward courtiers and their clients with monopoly licenses was regarded as outrageous. When Parliament abolished them through the statute of monopolies of 1624, the value of new invention was exempted, but inventors were distrusted, at best seen as overambitious and unrealistic visionaries.

An application for a patent had to go through as many as ten offices, each requiring a fee. Before the reforms of 1852, the cost of an English patent amounted to over a hundred pounds, Scottish and Irish patents were just as expensive. The inventor might also have to pay for the services of a patent agent. Once the patent had been granted, the protection of an invention involved the expense and uncertainty of litigation, requiring the services of specialist lawyers. Lawyers enriched themselves through patent disputes more than inventors. Inventors had to fend off Luddites (people who were deprived of their livelihood by the invention) on the one hand and interlopers who infringed their patents on the other. Judges examined the specification of an invention with the utmost care; the slightest error, say the misspelling of a word, might invalidate the whole document. After more than 20 years of debate in 1852, Parliament legislated to make the process more transparent and accessible to inventors.

Other countries developed their own systems. In France, the patent law of 1791 enshrined the concept of the inventor's natural right to own his intellectual property. Inventions were scrutinized by a committee of the Paris Academy; for example Lazare Carnot took part in this process. In the United States, one of the first acts of the new republic was to establish a patent system that facilitated the registration and exploitation of the inventor's intellectual property; this was radically reformed after the Civil War. Switzerland abolished its patent system in 1850 but reintroduced it by stages in 1888 and 1907; the Netherlands abolished theirs in 1869, but restored it in 1912. Each of the German states had its own system.

I am not sure when the first biography of an engineer appeared. Perhaps it was a life of the French military engineer Vauban, who attracted early biographers because of his success in raising or resisting sieges. In Britain, it was not until Samuel Smiles started writing the life stories of the self-taught heroes of the Victorian age. He began with his pioneering

Life of George Stephenson, published in 1857, which led to his well-known *Self-Help* of 1859 (reprinted, see Smiles, 2002). When this proved to be a success he combined his earlier life of George Stephenson with lives of Brindley, Smeaton, Rennie and Telford in a three-volume work entitled *Lives of the Engineers*, first published in 1862 and in an extensively revised new edition in 1874. He followed this up with his *Industrial Biography* (Smiles, 1863), which is largely concerned with the development of iron-working and machine tools. In 1865, his *Lives of Boulton and Watt* was published and a separate *Life of Telford* followed soon afterwards (Smiles, 1867). After a period when he wrote about other subjects, he returned in 1885 to industrial biography by editing the autobiography of Nasmyth. Although Smiles' biographies contain much valuable information, they were heavily influenced by Victorian values. Moreover, he highlighted certain individuals, placing others in the background who had an equally good claim to fame. Also, he relied too much on the oral tradition, and neglected the theoreticians.

After Smiles, a number of books with titles like *Great Engineers* (e.g., Rolt, 1962b) were published, in which the same few famous names tend to appear again and again. The subjects were mainly British or worked in Britain, but *Grosse Ingenieure*, by Conrad Matschoss, first published in 1937 (English translation 1939), profiles a fairly international selection of almost 40 engineers up to the time of the First World War. Since the Second World War, L. T. C. Rolt, who was both an engineer himself and a gifted writer, has revisited the Victorian engineers. His *Victorian Engineering* of 1970 makes an excellent introduction to the subject. His well-researched and readable biographies of Telford, the Stephensons, Watt and the younger Brunel achieve a high standard of scholarship. The books of J. G. Crowther (1939, 1962) on the lives of eighteenth and nineteenth scientists can also be recommended. Among serial publications, the most useful is *Technology and Culture*, the Journal of the Society for the History of Technology, of which the first issue is dated 1959, but there are others, most notably *History and Technology*, and more specialized publications, such as the *Transactions of the Newcomen Society for the Study of the History of Engineering and Technology*, which also contain valuable material.

Bibliography

Abrahamson, A. (1995) *Zworykin, Pioneer of Television*. Urbana and Chicago: University of Illinois Press.

Aitken, H. G. J. (1976) *Syntony and Spark: The Origins of Radio*. New York: Wiley Interscience.

Allibone, T. E. (1980) *Dennis Gabor. Biographical Memoirs of the Fellows of the Royal Society* **26**:107–8.

Appleyard, R. (1930) *Pioneers of Electrical Communication*. London: Macmillan.

Appleyard, R. (1933) *Charles Parsons: His Life and Work*. London: Constable.

Armytage, W. H. G. (1961) *A Social History of Engineering*. London: Faber and Faber.

Ayrton, H. (2007) *The Electric Arc*. Whitefish, MT: Kessinger Publishing.

Babbage, C. (2007) *On the Economy of Machinery and Manufactures*. BiblioBazaar.

Babbage, C. (2009a) *Reflections on the Decline of Science in England, and on Some of Its Causes*. BiblioBazaar.

Babbage, C. (2009b) *Passages From The Life Of A Philosopher*. Read Books.

Bailey, M. R. (ed) (2003) *Robert Stephenson the Eminent Engineer*. Aldershot: Ashgate Publishing.

Bathurst, B. (1999) *The Lighthouse Stevensons*. London: Harper Collins.

Bell, S. P. (1975) *Biographical Index of British Engineers in the 19th Century*. New York: Garland.

Bergaust, E. (1976) *Wernher von Braun: The Authoritative and Definitive Biographical Profile of the Father of Modern Space Flight*. Washington: National Space Institute.

Blomfield, Sir R. (1938) *Sébastien le Prestre de Vauban (1633–1707)*. London: Methuen.

Bornecque, R. (1984) *Le France de Vauban*. Paris: Arthaud.

Boucher, C. T. G. (1963) *John Rennie 1761–1821: The Life and Work of a Great Engineer*. Manchester: Manchester University Press.

Boucher, C. T. G. (1968) *James Brindley Engineer 1716–1772*. Norwich: Goose.

Bradfield, C. (1993) *Thomas Telford's Temptation*. Cleobury Mortimer: M&M Baldwin.

Brightfield, M. F. (1961) The coming of the railroad to early Victorian England, as viewed by novels of the period. *Technology and Culture* **2**:45–72.

Bruce, R. V. (1973) *Bell: Alexander Graham Bell and the Conquest of Solitude*. London: Victor Gollancz.

Bryant, L. (1967) The origins of the four-stroke cycle. *Technology and Culture* **8**:178–98.

Bryant, L. (1976) The development of the diesel engine. *Technology and Culture* **17**:432–46.

Buchanan, A. (1983) The Great Eastern controversy; a comment. *Technology and Culture* **24**:98–106.

Buchanan, R. A. (1989) *The Engineers: A History of the Engineering Profession in Great Britain, 1750–1914*. London: Jessica Kingsley Publishers.

Buchanan, A. (2002) *Brunel: The Life and Times of Isambard Kingdom Brunel*. London and New York: Hambledon Continuum.

Burton, A. (2000) *Richard Trevithick Giant of Steam*. London: Aurum Press.

Cardwell, D. S. L. (1965) Power technologies and the advancement of science, 1700–1825. *Technology and Culture* **6**:188–207.

Cardwell, D. S. L. (1978) Science and technology. *Technology and Culture* **17**:674–87.

Carnot, S. (1986) *Reflections on the Motive Power of Fire* (trans. and ed. by R. Fox). Manchester: Manchester University Press.

Church, W. C. (1892) *The Life of John Ericsson*. London: Sampson, Low, Marston and Company.

Clark, R. C. (1977) *Edison: The Man Who Made the Future*. London: Macdonald and Jane's.

Clements, P. (1970) *Marc Isambard Brunel*. London: Longman.

Clerk, D. (1896) *The Gas and Oil Engine*. London: Longmans Green.

Conot, R. (1979) *Thomas A. Edison: A Streak of Luck*. New York: Seaview Books.

Cooper, C. C. (1984) The Portsmouth system of manufacture. *Technology and Culture* **25**:182–225.

Cooper, C. C. (1991) Making inventions patent. *Technology and Culture* **32**:837–84.

Coulson, T. (1950) *Joseph Henry; his Life and Work*. Princeton, NJ: Princeton University Press.

Crowther, J. G. (1939) *Six Great Engineers*. London: H. Hamilton.

Crowther, J. G. (1962) *Scientists of the Industrial Revolution*. London: Cresset Press.

Davenport, W. W. (1978) *Gyro! The Life and Times of Lawrence Sperry*. New York: Charles Scribner's sons.

Davies, H. (1975) *George Stephenson: A Biographical Study of the Father of Railways*. London: Weidenfeld and Nicolson.

Dickinson, H. W. (1935) *James Watt: Craftsman and Engineer*. Cambridge: Cambridge University Press.

Dickinson, H. W. and Jenkin, R. (1927) *James Watt and the Steam Engine*. Oxford: Clarendon Press.

Dickinson, H. W. and Titley, A. (1934) *Richard Trevithick: The Engineer and the Man*. Cambridge: Cambridge University Press.

Diesel, R. (2007) *Solidarismus*. Augsburg: Maro Verlag.

Dornberger, W. (1963) The German V-2. *Technology and Culture* **4**:393–409.

Dorsey, F. L. (1947) *Road to the Sea: The Story of James B. Eads and the Mississippi River*. New York: Rinehart.

Drummond, C. (1994) *The Remarkable Life of Victoria Drummond: Marine Engineer*. London: The Institute of Marine Engineers.

Dunlap, O. S. (1937) *Marconi: The Man and his Wireless*. London: Macmillan.

Dunsheath, P. (1967) *Giants of Electricity*. New York: Thomas Y. Crowell.

Dutton, H. I. (1984) *The Patent System and Inventive Activity During the Industrial Revolution 1750–1852*. Manchester: Manchester University Press.

Emmerson G. S. (1977) *John Scott Russell: A Great Victorian Engineer and Naval Architect*. London: John Murray.

Emmerson, G. S. (1980) L. T. S. Rolt and the Great Eastern affair of Brunel versus Scott Russell. *Technology and Culture* **21**:553–69.

Evans, F. T. (1981) Roads, railways and canals. *Technology and Culture* **22**:1–34.

Ewing, J. A. (1931) The Hon. Sir Charles Parsons OM, KCB 1854–1931. *Proceedings of the Royal Society of London, Series A (Obituary Notices)* **131**:v–xxv.

Fairlie, G. and Cayley, E. (1965) *The Life of a Genius*. London: Hodder and Stoughton.

Feibleman, J. K. (1961) Pure science, applied science, technology, engineering: an attempt at definitions. *Technology and Culture* **2**:305–17.

Feilden, G. B. R. and Hawthorne, W. (1998) Sir Frank Whittle OM, KBE. *Biographical Memoirs of the Fellows of the Royal Society* **44**:435–52.

Flood, R., McCartney, M. and Whitaker, A. (2008) *Kelvin, Life, Labours and Legacy*. Oxford: Oxford University Press.

Fouche, R. (2003) *Black Inventors in the Age of Segregation*. Baltimore: Johns Hopkins University Press.

Garliński, J. (1978) *Hitler's Last Weapons: The Underground War Against the V1 and V2*. London: Friedman.

Geise, J. (1959) What is a railway? *Technology and Culture* **1**:68–77.

Gibb, Sir A. (1935) *The Story of Telford, the Rise of Civil Engineering*. London: A&C Black.

Gibbs-Smith, C. H. (1962) *Sir George Cayley's Aeronautics, 1796–1855*. London: Her Majesty's Stationery Office.

Gies, J. (1963) *Bridges and Men*. London: Cassell.

Gilfillan, S. C. (1935) *Sociology of Invention*. Cambridge, MA: MIT Press.

Gillispie, C. C. (1971) *Lazare Carnot Savant*. Princeton, NJ: Princeton University Press.

Goff, A. C. (1946) *Women Can be Engineers*. Ann Arbor, MI: Edwards Bros.

Golley, J. (1987) *Whittle: True Story*. Shrewsbury: Airlife.

Gordon, R. B. (1985) Hydrological science and the development of waterpower for manufacturing. *Technology and Culture* **26**:204–35.

Graham, L. R. (1993) *The Ghost of the Executed Engineer: Technology and the Fall of the Soviet Union*. Cambridge, MA: Harvard University Press.

Habakkuk, H. J. (1962) *American and British Technology in the Nineteenth Century: The Search for Labour-Saving Inventions*. Cambridge: Cambridge University Press.

Hacker, S. (1990) *Doing it the Hard Way*. Boston: Unwin Hyman.

Hadfield, C. and Skempton, A. W. (1979) *William Jessop, Engineer*. London: David and Charles.

Halliday, S. (1999) *The Great Stink of London: Sir Joseph Bazalgette and the Cleansing of the Victorian Capital*. Thrupp, Stroud: Sutton Publishing.

Harford, J. (1997) *Korolev*. New York: John Wiley.

Harrod, K. (1958) *Master Bridge Builders: The Story of the Roeblings*. New York: Julian Messner.

Hart, I. B. (1958) *James Watt and the History of Steam Power*. London: Abelard-Schuman.

Harvie, D. (2004) *Eiffel: The Genius who Reinvented Himself*. Stroud: Sutton.

Hendricks, G. (1961) *The Edison Motion Picture Myth*. Berkeley and Los Angeles: University of California Press.

Hertz, J. (ed.) (1977) *Hertz, Heinrich Memoirs, Letters, Diaries*. San Francisco, CA: San Francisco Press.

Hilaire-Perez, L. (1991) Invention and the state in eighteenth century France. *Technology and Culture* **32**:911–31.

Hong, S. (1994) Marconi and the Maxwellians. *Technology and Culture* **35**:717–49.

Houndhell, D. A. (1975) Elisha Gray and the telephone. *Technology and Culture* **16**:133–61.

How, L. (1900) *James B. Eads*. Boston: Houghton, Mifflin & Co.

Hughes, T. P. (1971) *Elmer Sperry; Inventor and Engineer*. Baltimore: Johns Hopkins Press.

Hunley I. D. (1985) The enigma of Robert H Goddard. *Technology and Culture* **36**:327–50.

Hunt, I. and Draper, W. (1964) *Lightning in his Hand: The Life Story of Nikola Tesla*. Sage Books.

Hyman, A. (1982) *Charles Babbage: Pioneer of the computer*. Oxford: Oxford University Press.

Israel, P. (1998) *Edison: A Life of Invention*. New York: John Wiley.

James, I. (2004) *Remarkable Physicists*. Cambridge: Cambridge University Press.

James, I. (2009a) *Driven to Innovate*. Oxford: Peter Lang.

James, I. (2009b) *Biographical Memoirs of the Fellows of the Royal Society* **55**:257–65.

Jarvis, A. (1997) *Samuel Smiles and the Construction of Victorian Values*. Thrupp, Stroud: Sutton Publishing.

Jolly, W. P. (1972) *Marconi*. London: Constable.

Joravsky, D. (1961) The history of technology in Soviet Russia and Marxist doctrine. *Technology and Culture* **2**:5–10.

Josephson, M. (1959) *Edison: A Biography*. New York: McGraw Hill.

Josephson, P. R. (1995) 'Projects of the century' in Soviet history. *Technology and Culture* **36**:519–59.

Kamm, T. and Baird, M. (2002) *John Logie Baird – A Life*. Edinburgh: National Museums of Scotland Publishing.

Kanefsky, J. and Robey, J. (1980) Steam engines in eighteenth century Britain. *Technology and Culture* **21**:167–86.

Kelly, F. C. (1944) *The Wright Brothers: A Biography Authorized by Orville Wright*. London: G. Harrap.

Kerber, L. L. (1996) *Stalin's Aviation Gulag a Memoir of Andrei Tupolev and the Purge Era*. (ed. by Von Hardesty). Washington DC: Smithsonian Institution Press.

Kerker, M. (1961) Science and the steam engine. *Technology and Culture* **2**:381–90.

King, A. G. (1925) *Kelvin the Man*. London: Hodder and Stoughton.

Kingsford, P. (1960) *F. W. Lanchester: The Life of an Engineer*. London: E. Arnold.

Kouwenhoven, J. A. (1982) The designing of the Eads bridge. *Technology and Culture* **23**:535–68.

Kurylo F. and Susskind, C. (1981) *Ferdinand Braun: A Life of the Nobel Prizewinner and Inventor of the Cathode Ray Oscilloscope*. Cambridge, MA: MIT Press.

Lanchester, F. W. (2009) *Aircraft in Warfare, the Dawn of the Fourth Arm*. BiblioBazaar.

Lasby, C. G. (1971) *Project Paperclip: German Scientists and the Cold War*. New York: Atheneum.

Lehman, M. (1988) *Robert H. Goddard, Pioneer of Space Research*. New York: De Capo Press.

Longmate, N. (1985) *Hitler's Rockets: The Story of the V-2s*. London: Atheneum.

Lubar, S. (1991) The transformation of antebellum patent law. *Technology and Culture* **32**:932–59.

McArthur, T. and Waddell, P. (1990) *Vision Warrior: The Hidden Achievement of John Logie Baird*. Orkney: Scottish Falcon.

Macleod, C. (1991) The paradoxes of patenting. *Technology and Culture* **32**: 885–911.

Macleod, C. (2007) *Heroes of Invention: Technology, Liberalism and British Identity*. Cambridge: Cambridge University Press.

MacMahon, J. R. (1930) *The Wright Brothers: Fathers of Flight*. Boston: Little, Brown & Co.

Malet, H. (1961) *The Canal Duke*. London: David and Charles.

Marconi, D. (1962) *My Father Marconi*. London: Frederick Muller.

Mason, J. (1991) Hertha Ayrton and the admission of Women to the Royal Society. *Notes and Records of the Royal Society* **45**:201–20.

Meynell, L. (1956) *James Brindley: The Pioneer of Canals*. London: W. Laurie.

Meynell, L. (1957) *Thomas Telford*. London: The Bodley Head.

Millard, A. (1990) *Edison and the Business of Invention*. Baltimore, MD: Johns Hopkins University Press.

Moore, D. L. (1977) *Ada, Countess of Lovelace: Byron's Legitimate Daughter*. London: John Murray.

Mosely, M. (1964) *Irascible Genius*. London: Hutchinson.

Moyer, A. E. (1997) *Joseph Henry: The Life of an American Scientist*. Washington DC: Smithsonian Institution Press.

Mukerji, C. (2009) *Impossible Engineering: Technology and Territoriality on the Canal du Midi*. Princeton, NJ: Princeton University Press.

Musson, A. E. and Robinson, E. (1969) *James Watt and the Steam Revolution*. London: Adams and Dart.

Nahum, A. (2004) *Frank Whittle: Invention of the Jet*. Duxford, Cambridge: Icon Books UK.

Needham, J. et al. (1954–2004) *Science and Civilisation in China*. Seven vols. Cambridge: Cambridge University Press.

Neufeld, M. J. (1990) Weimar culture and futuristic technology. *Technology and Culture* **31**:725–52.

Neufeld, M. J. (1993) *The Rocket and the Reich*. New York: The Free Press.

Nitske W. R. and Wilson, M. W. (1965) *Rudolf Diesel: Pioneer of the Age of Power*. Norman, OK: University of Oklahoma Press.

Noble, C. B. (1938) *The Brunels: Father and Son*. London: R. Cobden Sanderson.

O'Neill, J. (1981) *Prodigal Genius: The Life of Nikola Tesla*. Hollywood, CA: Angriff Press.

Panel, J. P. M. (1964) *An Illustrated History of Civil Engineering*. London: Thames and Hudson.

Picon, A. (1992) *French Architects and Engineers in the Age of Enlightenment*. (trans. by M. Thom) Cambridge: Cambridge University Press.

Pritchard, J. L. (1955) Sir George Cayley, Bart, the father of British Aeronautics: the man and his work. *The Journal of the Royal Aeronautical Society* **59**: 79–119.

Pritchard, J. L. (1961) *Sir George Cayley: The Inventor of the Aeroplane*. London: Parrish.

Prout, H. G. (1922) *A Life of George Westinghouse*. London: Benn Bros.

Pudney, J. (1974) *Brunel and his World*. London: Thames and Hudson.

Rae, J. B. (1961) Science and engineering in the history of aviation. *Technology and Culture* **2**:391–9.

Reyburn, W. (1972) *Bridge Across the Atlantic: The Story of John Rennie*. London: Harrap.

Robinson, E. (1972) James Watt and the law of patents. *Technology and Culture* **13**:115–39.

Rolt, L. T. C. (1957) *Isambard Kingdom Brunel*. London: Longmans Green.

Rolt, L. T. C. (1958) *Thomas Telford*. London: Longman.

Rolt, L. T. C. (1960) *George and Robert Stephenson*. London: Longman.

Rolt, L. T. C. (1962a) *James Watt*. London: Batsford.

Rolt, L. T. C. (1962b) *Great Engineers*. London: G. Bell and Sons.

Rolt, L. T. C. (1970) *Victorian Engineering*. London: Allen Lane.

Rolt, L. T. C. (1973) *From Sea to Sea*. London: Allen Lane.

Rolt, L. T. C. (2006) *Isambard Kingdom Brunel*. London: Penguin.

Rossman, J. (1964) *Industrial Creativity: The Psychology of the Inventor* (new edn). New Hyde Park, NY: University Books.

Rowland, J. (1954) *George Stephenson* London: Odhams.

Ruddock, T. (1979) *Arch Bridges and their Builders 1735–1835*. Cambridge: Cambridge University Press.

Scaife, W. G. (2000) *From galaxies to turbines : science, technology and the Parsons family*. Bristol: Institute of Physics.

Scherer, F. M. (1965) Invention and innovation in the Watt–Boulton steam-engine venture. *Technology and Culture* **6**:165–87.

Schuyler, H. (1931) *The Roeblings: A Century of Engineers, Bridge Builders and Industrialists*. Princeton, NJ: Princeton University Press.

Seifer, M. J. (1998) *Wizard: The Life and Times of Nikola Tesla; Biography of a Genius*. Secaucus, NJ: Citadel Press.

Sharlin, H. I., in collaboration with Sharlin, T. (1979) *Lord Kelvin, the Dynamic Victorian*. University Park and London: Pennsylvania State University Press.

Sharp, E. (1926) *Hertha Ayrton 1854–1923*. London: Edward Arnold.

Shurkin, J. N. (2006) *Broken Genius: The Rise and Fall of William Shockley, Creator of the Electronic Age*. Basingstoke: Macmillan.

Siemens, W. von (1966) *Inventor and Entrepreneur; Recollections of Werner von Siemens*. London: Lund and Humphries.

Simonds, W. A. (1935) *Edison – his Life, his Work, his Genius*. Brooklyn, NY: Bobbs-Merrill Co.

Singer, C. J., Williams, T. I. and Raper, R. (1954–84) *A History of Technology*. vols 1–6. Oxford: Oxford University Press.

Skeat, W. O. (1973) *George Stephenson: The Engineer and his Letters*. London: The Institution of Mechanical Engineers.

Sloane, N. J. A. and Wyner, A. D. (eds) (1993) *Claude Elwood Shannon: Collected Papers*. New York: IEE Free Press.

Smeaton, J. (1791) *A Narrative of the Building, and a Description of the Construction of the Edystone Lighthouse ... To Which is Subjoined an Appendix, Giving some Account of the Lighthouse on the Spurn Point*. London: G. Nichol.

Smeaton, J. (1938) *Diary of his Journey to the Low Countries* 1755 (intro. by Arthur Tilley). Leamington Spa: Courier Press.

Smiles, S. (1857) *The Life of George Stephenson*. London: John Murray.

Smiles, S. (1862, new edn 1874) *Lives of the Engineers*. London: John Murray.

Smiles, S. (1863) *Industrial Biography: Iron Workers and Tool Makers*. London: John Murray.

Smiles S. (1865) *Lives of Boulton and Watt*. London: John Murray.

Smiles, S. (1867) *Life of Telford*. London: John Murray.

Smiles, S. (ed.) (1885) *James Nasmyth, Engineer: An Autobiography*. London: John Murray.

Smiles, S. (2002) *Self-Help*. Oxford: Oxford University Press.

Smith, C. and Wise, M. N. (1989) *Energy and Empire: A Biographical Study of Lord Kelvin*. Cambridge: Cambridge University Press.

Smith, D. (1986/7) Sir Joseph William Bazalgette (1819–1891): engineer to the Metropolitian Board of Works. *Transactions of the Newcomen Society* **58**:89–112.

Smith, D. (ed.) (1994) *Perceptions of Great Engineers*. London: Science Museum.

Stein, D. (1985) *Ada: A Life and a Legacy*. Cambridge, MA: MIT Press.

Steiner, F. H. (1981) Building with iron. *Technology and Culture* **22**:700–24.

Steinman, D. B. (1950) *The Builders of the Bridge: The Story of John Roebling and his Son*. New York: Harcourt Brace.

Sullivan, O. R. (1998) *African-American Inventors*. New York: John Wiley and Sons.

Thomas, D. E. (1978) Diesel Father and Son. *Technology and Culture* **19**:376–93.

Thomas, D. E. (1987) *Diesel: Technology and Society in Industrial Germany*. Tuscaloosa, AL: University of Alabama Press.

Toole, B. A. (1992) *The Enchantress of Numbers*. Mill Valley, CA: Strawberry Press.

Turner, T. and Skempton, A. W. (1981) John Smeaton. In *John Smeaton FRS* (ed. by A. W. Skempton). London: Thomas Telford, Ltd.

Vauban, S. (1740) *Memoir pour servir a l'instruction dans la conduite des sièges.* Leiden: Jean & Herman Verbeek.

Vignoles, K. H. (1962) *Charles Black Vignoles: Romantic Engineer*. Cambridge: Cambridge University Press.

Wachhorst, W. (1981) *Thomas Alva Edison: An American Myth.* Cambridge, MA: MIT Press.

Ward, R. J. (2006) *From Nazis to NASA*. Stroud: Sutton Publishing.

Warson, S. J. (1954) *Carnot*. London: The Bodley Head.

Woolley, B. (1999) *The Bride of Science*. London: Macmillan.

Yager, R. (1968) *James Buchanan Eads: Master of the Great River*. Princeton, NJ: van Nostrand.

Young, R. A. (1978) *The Flying Bomb*. London: Ian Allen.

Credits

Riquet: Rolt 1973; Mukerji 2009
Vauban: Blomfield 1938; Bornecque 1984
Brindley: Smiles 1862; Meynell 1956; Boucher 1968
Smeaton: Smeaton 1938; Hart 1958; Musson and Robinson 1969; Turner and Skempton 1981
Watt: Smiles 1865; Dickinson and Jenkin 1927; Dickinson 1935; Rolt 1962a
Jessop: Hadfield and Skempton 1979; Bradfield 1993
Carnot, Sr: Gillispie 1971
Telford: Smiles 1865; Gibb 1935; Meynell 1957; Rolt 1958; Bradfield 1993
Rennie: Boucher 1963; Reyburn 1972
Brunel, Sr: Noble 1938; Clements 1970
Trevithick: Dickinson and Titley 1954; Burton 2000
Cayley: Pritchard 1961; Gibbs-Smith 1962
Stephenson, Sr: Rowland 1954; Rolt 1960; Skeat 1973; Davies 1975
Babbage: Mosely 1964; Bell 1975; Hyman 1988
Henry: Coulson 1950; Moyer 1997
Vignoles: Vignoles 1982
Carnot, Jr: Gillispie 1971; Carnot 1986
Ericsson: Church 1982
Stephenson, Jr: Rolt 1960; Bailey 2003
Brunel, Jr: Noble 1938; Rolt 1957; Pudney 1974; Emmerson 1977; Buchanan 1983
Roebling: Schuyler 1931; Steinman 1950; Harrod 1958
Bazalgette: Yager 1968; Smith 1986/7; Halliday 1999
Eads: How 1900; Dorsey 1947; Kouwenhaven 1982
Thomson: King 1925; Smith and Wise 1989; James 2004; Flood *et al.* 2008
Eiffel: Harvie 2004
Westinghouse: Prout 1922
Edison: Simonds 1934; Josephson 1959; Hendricks 1961; Clark 1977; Conot 1979; Wachhorst 1981; Millard 1990; Israel 1998
Bell: Bruce 1973, Houndhell 1975
Braun: Kurylo and Susskind 1981
Ayrton: Sharp 1926; Mason 1991; James 2009a
Parsons: Appleyard 1933; Scaife 2000
Woods: Fouché 2003
Tesla: Hunt and Draper 1964; Fairlie and Cayley 1965; O'Neill 1981; Seifer 1998
Hertz: Appleyard 1930; James 2009a

Diesel: Nitske and Wilson 1965; Bryant 1976; Thomas 1978; 1987
Sperry: Hughes 1971; Davenport 1978
Wrights: McMahon 1930; Kelly 1944
Lanchester: Kingsford 1960
Marconi: Dunlap 1937; Marconi 1962; Jolly 1972; Aitken 1976; Hong 1994
Pal'chinskii: Graham 1993
Clarke: Goff 1946
Tupolev: Kerber 1996
Baird: McArthur and Waddell 1990; Kamm and Baird 2002
Zworykin: Abrahamson 1995
Gabor: Allibone 1980; James 2009a
Korolev: Harford 1997
Whittle: Golley 1987; Feilden and Hawthorne 1998; Nahum 2004
Shockley: Shurkin 2006
Von Braun: Dornberger 1963; Lasby 1971; Bergaust 1976; Garliński 1978; Young 1978; Longmate 1985; Neufeld 1993; Ward 2006
Shannon: Sloane and Wyner 1993; James 2009b

Image credits

Charles Babbage; John Logie Baird; Sadi Carnot, George Cayley; Lazare Carnot; Rudolf Diesel; John Ericsson; William Jessop; Sergei Korolyev; Frederick Lanchester; Guglielmo Marconi; John Rennie; Richard Trevithick; Thomas Telford; Frank Whittle; Wilbur and Orville Wright; James Watt; Vladimir Zworykin: courtesy of Science Photo Library

Edith Clark: courtesy of Schenectady Museum & Suits-Bueche Planetarium

Alexander Graham Bell; Joseph Henry; Elmer Sperry; George Westinghouse: courtesy of the Library of Congress/Science Photo Library

Isambard Kingdom Brunel: © Hulton-Deutsch Collection/CORBIS

Marc Isambard Brunel: painting by S. Drummond, courtesy of Science Photo Library

Sir Joseph William Bazalgette: © Hulton-Deutsch Collection/CORBIS

Karl Ferdinand Braun: courtesy of the National Library of Congress/Science Photo Library

James Brindley; Robert Stephenson: courtesy George Bernard, Science Photo Library

Thomas Edison: courtesy of John Daugherty, Science Photo Library

Gustav Eiffel: © Bettmann/CORBIS

Dennis Gabor; William Shockley: courtesy of Emilio Segre Visual Archives/American Institute of Physics/Science Photo Library

Heinrich Hertz: © Bettmann/CORBIS

Charles Parsons: courtesy of George Grantham Bain Collection/Library of Congress/Science Photo Library

John Smeaton; Charles Vignoles: courtesy of the Royal Astronomical Society/Science Photo Library

George Stephenson: courtesy of Ken Welsh, The Bridgeman Art Library

Niokla Tesla: courtesy of the USA Library of Congress/Science Photo Library

Andrei Tupolev: courtesy of Ria Novosti, Science Photo Library

Sebastien le Prestre de Vauban: © Bettmann/CORBIS

Wernher von Braun: courtesy of NASA/Science Photo Library

Granville Woods: courtesy of the Schomburg Center for Research in Black Culture/New York Public Library/Science Photo Library